AF361902

Carbon and Nitrogen in Forest Ecosystems—Series I

Carbon and Nitrogen in Forest Ecosystems—Series I

Editor

Yowhan Son

MDPI • Basel • Beijing • Wuhan • Barcelona • Belgrade • Manchester • Tokyo • Cluj • Tianjin

Editor
Yowhan Son
Korea University
Korea

Editorial Office
MDPI
St. Alban-Anlage 66
4052 Basel, Switzerland

This is a reprint of articles from the Special Issue published online in the open access journal *Forests* (ISSN 1999-4907) (available at: https://www.mdpi.com/journal/forests/special_issues/Carbon_Nitr_system).

For citation purposes, cite each article independently as indicated on the article page online and as indicated below:

LastName, A.A.; LastName, B.B.; LastName, C.C. Article Title. *Journal Name* **Year**, *Article Number*, Page Range.

ISBN 978-3-03936-744-3 (Hbk)
ISBN 978-3-03936-745-0 (PDF)

Cover image courtesy of Yowhan Son.

Contents

About the Editor . . . . **vii**

Preface to "Carbon and Nitrogen in Forest Ecosystems—Series I" . . . . **ix**

Jie Yuan, Shibu Jose, Zhaoyong Hu, Junzhu Pang, Lin Hou and Shuoxin Zhang
Biometric and Eddy Covariance Methods for Examining the Carbon Balance of a
Larix principis-rupprechtii Forest in the Qinling Mountains, China
Reprinted from: *Forests* **2018**, *9*, 67, doi:10.3390/f9020067 . . . . **1**

Shu Fang, Zhibin He, Jun Du, Longfei Chen, Pengfei Lin and Minmin Zhao
Carbon Mass Change and Its Drivers in a Boreal Coniferous Forest in the Qilian Mountains,
China from 1964 to 2013
Reprinted from: *Forests* **2018**, *9*, 57, doi:10.3390/f9020057 . . . . **25**

Matthew Powers, Randall Kolka, John Bradford, Brian Palik and Martin Jurgensen
Forest Floor and Mineral Soil Respiration Rates in a Northern Minnesota Red
Pine Chronosequence
Reprinted from: *Forests* **2018**, *9*, 16, doi:10.3390/f9010016 . . . . **41**

Pablo I. Fragoso-López, Rodrigo Rodríguez-Laguna, Elena M. Otazo-Sánchez,
César A. González-Ramírez, José René Valdez-Lazalde, Hermann J. Cortés-Blobaum
and Ramón Razo-Zárate
Carbon Sequestration in Protected Areas: A Case Study of an *Abies religiosa* (H.B.K.)
Schlecht. et Cham Forest
Reprinted from: *Forests* **2017**, *8*, 429, doi:10.3390/f8110429 . . . . **57**

Jian Yang, Xin Chang Zhang, Zhao Hui Luo and Xi Jun Yu
Nonlinear Variations of Net Primary Productivity
and Its Relationship with Climate and Vegetation Phenology, China
Reprinted from: *Forests* **2017**, *8*, 361, doi:10.3390/f8100361 . . . . **71**

Qiong Wang, Fayun Li, Xiangmin Rong and Zhiping Fan
Plant-Soil Properties Associated with Nitrogen Mineralization: Effect of Conversion of Natural
Secondary Forests to Larch Plantations in a Headwater Catchment in Northeast China
Reprinted from: *Forests* **2018**, *9*, 386, doi:10.3390/f9070386 . . . . **93**

Cole D. Gross, Jason N. James, Eric C. Turnblom and Robert B. Harrison
Thinning Treatments Reduce Deep Soil Carbon and Nitrogen Stocks in a Coastal Pacific
Northwest Forest
Reprinted from: *Forests* **2018**, *9*, 238, doi:10.3390/f9050238 . . . . **109**

Chen Ning, Gregory M. Mueller, Louise M. Egerton-Warburton, Andrew W. Wilson,
Wende Yan and Wenhua Xiang
Diversity and Enzyme Activity of Ectomycorrhizal Fungal Communities Following Nitrogen
Fertilization in an Urban-Adjacent Pine Plantation
Reprinted from: *Forests* **2018**, *9*, 99, doi:10.3390/f9030099 . . . . **129**

Joseph E. Knelman, Emily B. Graham, Scott Ferrenberg, Aurélien Lecoeuvre,
Amanda Labrado, John L. Darcy, Diana R. Nemergut and Steven K. Schmidt
Rapid Shifts in Soil Nutrients and Decomposition Enzyme Activity in Early Succession
Following Forest Fire
Reprinted from: *Forests* **2017**, *8*, 347, doi:10.3390/f8090347 . . . . **147**

Zhaofeng Lei, Huanfa Sun, Quan Li, Junbo Zhang and Xinzhang Song
Effects of Nitrogen Deposition on Soil Dissolved Organic Carbon and Nitrogen in Moso Bamboo
Plantations Strongly Depend on Management Practices
Reprinted from: *Forests* **2017**, *8*, 452, doi:10.3390/f8110452 . . . . **159**

About the Editor

Yowhan Son is mainly interested in carbon and nutrient distribution and cycling in forest ecosystems in addition to in the context of natural and human influences on ecosystem structure and function, organic matter production and decomposition in soils, nitrogen cycling in trees and forest soils, and soil and water pollution under different natural and human disturbances in forests. Prof. Son is an Editorial Advisory Board Member of *Forest Ecology and Management*.

Preface to "Carbon and Nitrogen in Forest Ecosystems—Series I"

When studying forest ecosystems, it is essential to understand the differences between carbon and nitrogen spatial and temporary distribution and cycling when using various approaches. In addition, biotic/abiotic factors and natural/artificial disturbances on carbon and nitrogen cycling need to be better understood to draw implication on forest management practices.

Relevant matters investigated within this Special Issue are as follows:

- different approaches to measure carbon and nitrogen distribution and cycling in forest ecosystems including field measurement, remote sensing, and modeling;

- differences in carbon and nitrogen cycling within an ecosystem and among ecosystems;

- changes in carbon and nitrogen cycling in forest ecosystems along successional gradients;

- roles of microbes, insects, and animals in carbon and nitrogen cycling in forest ecosystems;

- influences of climate change on carbon and nitrogen cycling in forest ecosystems;

- artificial manipulation of trees to simulate carbon and nitrogen cycling to climate change;

- influences of forest management practices on carbon and nitrogen cycling in forest ecosystems;

- ecosystem based forest management.

This Special Issue aims to understand carbon and nitrogen distribution and cycling in forest ecosystem for ecosystem-based forest management under different natural and artificial disturbances.

Yowhan Son
Editor

Article

Biometric and Eddy Covariance Methods for Examining the Carbon Balance of a *Larix principis-rupprechtii* Forest in the Qinling Mountains, China

Jie Yuan [1], Shibu Jose [2], Zhaoyong Hu [1], Junzhu Pang [1,3], Lin Hou [1,3] and Shuoxin Zhang [1,3,*]

[1] College of Forestry, Northwest A&F University, Xianyang 712100, China; yuanjie@nwsuaf.edu.cn (J.Y.); huzy418@gmail.com (Z.H.); pangjunzhu@nwsuaf.edu.cn (J.P.); houlin1969@163.com (L.H.)
[2] School of Natural Resources, University of Missouri, Columbia, MO 65211, USA; joses@missouri.edu
[3] Qinling National Forest Ecosystem Research Station, Huoditang, Ningshan 711600, China
* Correspondence: sxzhang@nwsuaf.edu.cn; Tel./Fax: +86-29-8708-2993

Received: 20 December 2017; Accepted: 25 January 2018; Published: 29 January 2018

Abstract: The carbon balance of forests is controlled by many component processes of carbon acquisition and carbon loss and depends on the age of vegetation, soils, species composition, and the local climate. Thus, examining the carbon balance of different forests around the world is necessary to understand the global carbon balance. Nevertheless, the available information on the carbon balance of *Larix principis-rupprechtii* forests in the Qinling Mountains remains considerably limited. We provide the first set of results (2010–2013) from a long-term project measuring forest-atmosphere exchanges of CO_2 at the Qinling National Forest Ecosystem Research Station (QNFERS), and compare the net ecosystem exchange (NEE) based on biometric measurements with those observed via the eddy covariance method. We also compare the total ecosystem respiration via scaled-up chamber and eddy covariance measurements. The net primary productivity (NPP) was 817.16 ± 81.48 g·C·m^{-2}·y^{-1}, of which ΔB_{living} and D_{total} accounted for 77.7%, and 22.3%, respectively. Total ecosystem respiration was 814.47 ± 64.22 g·C·m^{-2}·y^{-1}, and cumulative annual soil respiration, coarse woody debris respiration, stem respiration, and leaf respiration were 715.47 ± 28.48, 15.41 ± 1.72, 35.28 ± 4.78, and 48.31 ± 5.24 g·C·m^{-2}·y^{-1}, respectively, accounting for 87.85%, 1.89%, 4.33%, and 5.93% of the total ecosystem respiration. A comparison between ecosystem respiration from chamber measurements and that from eddy covariance measurements showed a strong linear correlation between the two methods ($R^2 = 0.93$). The NEE of CO_2 between forests and the atmosphere measured by eddy covariance was -288.33 ± 25.26 g·C·m^{-2}·y^{-1}, which revealed a carbon sink in the *L. principis-rupprechtii* forest. This number was 14% higher than the result from the biometric measurements (-336.71 ± 25.15 g·C·m^{-2}·y^{-1}). The study findings provided a cross-validation of the CO_2 exchange measured via biometric and eddy covariance, which are beneficial for obtaining the true ecosystem fluxes, and more accurately evaluating carbon budgets.

Keywords: carbon balance; Qinling Mountains; biomass regression model; eddy covariance; net primary productivity; net ecosystem exchange

---

1. Introduction

Forests play a critical role in the global carbon cycle [1,2], and since the 1990s, substantial data have been acquired to clarify the contributions of forest ecosystems to the global carbon cycle [3–5]. The Qinling Mountains in central China provide an important climate boundary between the southern subtropics and the northern temperate zone, where the typical vegetation of both climate zones is present together with astonishingly high biodiversity [6]. Nevertheless, information relevant to the carbon balance from forests in the Qinling Mountains remains considerably limited. Studies have demonstrated that mountain forests are 'hot spots' for carbon cycling and are expected to be more strongly affected by climate change than lowland forests due to their sensitivity to warming [7,8]. Therefore, there is an urgent need for increased knowledge about the carbon fluxes in various mountain forests, especially those in the Qinling Mountains.

Larix principis-rupprechtii is adapted to high light levels and can tolerate freezing temperatures. This species grows in deep, well-drained acidic or neutral soils and is a valuable reforestation species in China that is distributed over ten provinces. Due to its rapid growth, high-quality wood, resistance to adverse climate and soil conditions, and high wind resistance, the tree is used for forest regeneration and afforestation of barren hills. *L. principis-rupprechtii* forests in the Qinling Mountains serve as major research sites for forest ecosystem studies because they represent the regional vegetation in the temperate coniferous forest domain of China and are also a major component of temperate forests globally. In addition, the *L. principis-rupprechtii* forest is sensitive to global change [9]. In the Qinling Mountains, from 1958 to 1986, the area afforested with *Larix* reached 0.3×10^4 ha [10]. Although Zhou et al. estimated the carbon budget of a *Larix* forest in China [11], there is still considerable uncertainty about the strength of the carbon source/sink in this forest due to discrepancies in estimation methods and variations in age, management, and climate [12,13]. Thus, to accurately determine the carbon balance of *Larix* forests, an adequate understanding of the processes that control net CO_2 exchange in a young *L. principis-rupprechtii* planted forest in the temperate regions is required.

The carbon balance at the ecosystem level (net ecosystem exchange, NEE) is controlled by many component processes of carbon acquisition (photosynthesis, tree growth, forest ageing, and carbon accumulation in soils) and carbon loss (respiration of living biomass, tree mortality, microbial decomposition of litter, oxidation of soil carbon, degradation, and disturbance) [14]. However, previous studies have suggested respiration as the main determinant in controlling the carbon balance of ecosystems [13]. Ecosystem respiration is composed of autotrophic and heterotrophic components, whose contributions to total respiration vary in space and time. Components of respiration include soil (roots and microorganisms), coarse woody debris (CWD), and stem and leaf respiration, which are controlled by the complex interaction of many factors, including temperature, moisture, canopy cover, stand age, and nutrient contents [4,15]. Few studies have investigated the seasonal and annual variability of these respiratory components in detail [4,16]. Hence, it is absolutely necessary to quantify the ecosystem's respiratory components, which can allow researchers to determine the contribution of each component flux to the overall ecosystem respiration and improve our understanding of ecosystem respiration dynamics [17].

Currently, the foremost techniques for measuring NEE are the eddy covariance technique and the biometric technique. Each technique has advantages and disadvantages. The eddy covariance method is a micrometeorological technique and has been widely used in different ecosystems [18,19]. The eddy covariance technique has numerous advantages: (1) it is nondestructive and has a low workload, (2) it provides observations at the ecosystem scale, and (3) it yields continuous records that address time scales every half hour to the length of the data record [16,20]. Since the early 1990s, more than 500 eddy covariance flux towers have been built in numerous ecosystems around the world [21]. However, there are still deficiencies in the eddy covariance method. Firstly, the measurements become unreliable or unavailable when the atmospheric conditions (wind, temperature, humidity, CO_2) are unsteady, the terrain is uneven, or there is very weak turbulence, as sometimes occurs at night [22,23]. Secondly, this method is valid for usage on large-scale field plots and cannot provide information on the component carbon fluxes [16]. The biometric method offers the advantages of lower cost and simplicity, in principle. Moreover, the chambers are portable and well suited for small-scale studies, which is appropriate for replicated measurements in multiple small plots of field trials and is also necessary for estimating the contributions of component carbon fluxes (for example, net primary production (NPP), heterotrophic respiration, and autotrophic respiration) to the total fluxes. However, some drawbacks have limited the application of the biometric method for NEE measurements. For example, (1) a variety of potential errors, such as modifications in the enclosed microclimate, pressure artefacts, and spatial heterogeneity, may occur [24]; (2) it cannot effectively sample the full spatial variation of patch-specific fluxes; (3) it cannot observe the full short-term (intra-daily) and intermediate (inter-daily) temporal variations that occur within a site; and (4) large uncertainties in scaled-up estimates may result in over- or underestimates of the actual fluxes [25].

In short, both the eddy covariance and the biometric methods must be inaccurate in measuring NEE due to the weaknesses associated with using either method alone. Few studies have conducted comparisons of eddy covariance and biometric-based measurements of NEE, especially relative to measurements taken simultaneously at the same site using the two different methods [26]. Thus, comparing the NEE measured using the eddy covariance and the biometric methods is necessary to highlight the potential sources of errors. Such a comparison is straightforward but requires a strict methodology for testing the accuracy and consistency of the eddy covariance and the biometric fluxes.

This paper presents the first set of results (2010–2013) from a long-term project measuring forest-atmosphere exchanges of CO_2 using the eddy covariance and biometric methods in a *L. principis-rupprechtii* forest in the Qinling Mountains. The objectives of our study were as follows: (1) to describe measurements of soil, coarse woody debris (CWD), stem, and leaf respiration based on chamber methods and to combine these measured fluxes with continuous measurements of temperature to model the respiration of each ecosystem component; (2) to estimate the spatial and temporal variability of ecosystem respiration and the percentage of the total ecosystem respiration of each component based on chamber measurements; and (3) to compare the total ecosystem respiration based on scaled-up chamber measurements and NEE based on biometric measurements with those observed via the eddy covariance method and to evaluate the carbon balance of the *L. principis-rupprechtii* forest at this site.

2. Materials and Methods

2.1. Study Area

The study area for the ecosystem component measurements covered 1 ha centered on a tower equipped for eddy covariance measurements of carbon dioxide exchange located at the Huoditang Experimental Forest Farm of Northwest A&F University in the Qinling Mountains, Shaanxi Province, China (Figure S1). The altitude is 2150 m, and the geographic coordinates are 33°27′42″ N latitude and 108°28′54″ E longitude. The annual average temperature is 10.80 °C, the annual precipitation 1200 mm, and the climate belongs to the temperate zone. The period of snow cover is from December to March, with a maximum depth of approximately 20 cm. The soil is classified as mountain brown earth.

The study area was selectively logged in the 1960s and 1970s, and since then, there have been no major anthropogenic disturbances except for small amounts of illegal logging. Since the natural forest protection project was initiated in 1998, human activities have almost vanished in the region. To reduce disturbance, the permanent plot was protected by an enclosure. The site is level (a mean slope of 5°), which is ideal for this study, and the overstory and understory of the forest are homogeneous. Moreover, the results of the data quality estimate (footprint analysis, energy balance estimate, eddy statistic estimate, and power spectrum estimate) implied that not only the location selected but also the configuration of this observational system are comparable for observations of the fluxes in the long run [27].

The forest used for the current research was 50 years old and was dominated by *L. principis-rupprechtii*. The mean stand height, diameter at breast height (DBH), and stand density were 16 m, 18 cm, and 1585 trees·ha^{-1}, respectively. In the shrub layer, the height varied from 80 cm to 520 cm and the percent cover was 28%. The major shrubs species present were *Euonymus phellomanus*, *Lonicera hispida*, *Lindera glauca*, and *Rubus pungens*, together with herbs such as *Carex leucochlora*, *Deyeuxia arundinacea*, *Lysimachia christinae*, *Thalictrum minus*, *Anaphalis aureopunctata*, *Dioscorea nipponica*,

Rubia cordifolia, and *Sinacalia tangutica*, and the fern *Dryopteris goeringiana*. The average height of the herbs was 60 cm, and the percent cover was 40%.

2.2. Biometric Measurements

2.2.1. Plot Measurements

In summer 2009, we established a permanent plot in the *L. principis-rupprechtii* forest. The 1 ha plot was divided into 25 quadrats 20 m × 20 m in size. The quadrats were each subdivided into 16 sub-quadrats 5 m × 5 m in size. A total of 1585 trees and 2728 shrubs in these sub-quadrats were permanently marked with aluminum labels and numbered consecutively. Based on the plot investigation, 10 standard trees outside the plot were felled. The leaves and branches at different canopy positions and orientations and the stems of different diameters were all collected, and the roots were dug up from the 10 standard trees to measure the carbon content ratio and evaluate the biomass. The DBH of all trees (including dead and new trees) were documented in August of each year during 2009–2013 to estimate (the annual change in) the biomass, which was calculated via the regression model developed in a previous study in this region (Table S2) [28].

We also documented the species, height, crown width, and basal stem diameter of all shrubs (including dead and new shrubs) in August of each year during 2009–2013 for biomass calculations. Based on the species present within the plots, *E. phellomanus*, *L. hispida*, *L. glauca*, and *R. pungens* outside the plot were dug up, with totals of 55, 48, 78, and 62 individual plants, respectively. The species, height, crown width, and basal stem diameter of these harvested shrubs were recorded, and they were then taken back to the laboratory to measure the carbon content ratio and develop a biomass regression model (Table S3).

To reduce disturbance, based on the plot investigation, we selected twenty 1 m × 1 m groundcover quadrats outside the plot in August of 2009–2013. All of the herbs were dug out in the twenty quadrats each year in order to measure the carbon content ratio and biomass. The twenty quadrats were not repeated each year, and new quadrats were selected each year.

In order to accurately estimate the root biomass and correct the fine root loss caused by digging, we used soil coring to supplement the fine root biomass [29]. A representative root sample was extracted from soil cores of 30 cm in length and 1.8 cm in diameter (76 cm^3). We selected twenty soil cores outside the plot in August of 2009–2013. All soil cores were extracted each year to supplement the fine root biomass. The twenty soil cores were not repeated each year, and new soil cores were selected each year. Litterfall was collected from the beginning of August 2009 to August 2013 at monthly intervals. Twenty 1 m × 1 m litter traps were randomly erected in the plot. Each trap consisted of 2 mm mesh nylon netting (on a wooden frame) suspended from a wire hoop and held 30 cm above the ground by four metal poles.

2.2.2. Carbon Content Ratio

The samples of trees, shrubs, and herbs were classified into species and organs (stem, bark, branch, leaf, and root), and the stems, branches, and roots were cut into 10 cm lengths. In each species, the same organs of the samples were pooled into one composite sample, while the twenty litterfall traps were also pooled into one composite sample. These composite samples were dried at 85 °C to constant weight (approximately 72 h) and then crushed to pass through a No. 200 mesh (0.074 mm) in order to measure the carbon content ratio. Each composite sample was repeatedly measured three times with a TOC analyzer (TOC-VTH-2000A, Shimadzu, Japan), and the average value was obtained for the carbon content ratio of litterfall and the different organs of the trees, shrubs, and herbs.

2.2.3. Net Primary Productivity (NPP)

Forest NPP estimates have been based primarily upon measurements of stems, bark, branches, and roots (including coarse and fine roots) biomass gain using regression models for trees and shrubs and harvest methods for other ecosystem components (herbs and litterfall). NPP was estimated using the following equations:

$$NPP = \Delta B_{living} + D_{total} \tag{1}$$

$$\Delta B_{living} = \sum T_i O_i + \sum S_j O_j + \sum H_r O_r \tag{2}$$

$$D_{total} = \sum D_{ti} O_i + \sum D_{sj} O_j + L \times P \tag{3}$$

where ΔB_{living} is the increment in live plant biomass, T_i is the live tree biomass increment of the ith organ (except for leaf), O_i is the tree carbon content ratio of the ith organ, S_j is the live shrub biomass increment of the jth organ (except for leaf), O_j is the shrub carbon content ratio of the jth organ, H_r is the herbaceous biomass increment of the root, O_r is the herbaceous carbon content ratio of the root, D_{total} is the sum of dead plant mass, D_{ti} is the dead tree mass of the ith organ (except for leaf), O_i is the tree carbon content ratio of the ith organ, D_{sj} is the dead shrub mass of the jth organ (except for leaf), O_j is the shrub carbon content ratio of the jth organ, L is the mass of litterfall, and P is the carbon content ratio of litterfall. Herbivore loss is often assumed to be negligible in healthy stands [30] and was not estimated in this study.

2.2.4. Leaf Area Index

The LAI-2000 plant canopy analyzer (LI-COR, Inc., Lincoln, NE, USA) is designed to estimate the leaf area index (LAI) of plant canopies indirectly from measurements of radiation above and below the canopy based on a theoretical relationship between leaf area and canopy transmittance [31]. The below-canopy measurements were made at 40 points, which were marked with red stakes and located along permanent transects; the sampling distance was 15 m in this forest. Above-canopy measurements were taken automatically every 15 s by a second instrument in the center of an open field situated nearby. The fish-eye lens of the instrument was covered by a view cap with a 90° opening to ensure that the reference measurements were not influenced by the trees surrounding the clearings or by the operator [32]. In taking canopy measurements, the sensor was held so that the same portion of the sky and the same level (between 1 and 1.5 m above ground) was occluded for both sensors (above- and below-canopy measurements). The LAI measurements were made every 2 weeks from April to November in 2010–2013.

2.2.5. Micrometeorological Measurements

A full suite of micrometeorological measurements was taken from the weather station located 20 m away from the plot, including air temperature and humidity (HMP45C, Vaisala, Helsinki, Finland), photosynthetically active radiation, soil temperature (10 cm), and precipitation. Data from all the sensors were recorded on data loggers (CR-1000, Campbell Scientific, Logan, UT, USA), and the data were downloaded every 2 weeks to a laptop personal computer (PC).

2.2.6. Soil Respiration

Soil respiration was measured using an LI-6400-09 soil chamber connected to an LI-6400 portable photosynthesis system (LI-COR, Inc., Lincoln, NE, USA). Thirty soil collars, each with a height of 10 cm and a diameter of 10 cm, were randomly placed in the 1 ha plot. To avoid influence on the measurement of soil respiration, the soil collars were inserted into the soil at the depth of 2 cm one week before the measurement of soil respiration. The surface vegetation inside the soil collars was cleared 1 day before the measurement, and the topsoil was kept intact to avoid its influence on the measured results. Surface efflux was measured three times in succession for each collar during each measurement period. Soil temperature at 10 cm was measured adjacent to each respiration collar with a portable temperature probe provided with the LI-6400. The measurements were made every 2 weeks from April to November in 2010–2013.

We used an exponential equation to analyze the relationship between respiration and temperature:

$$R = R_0 e^{\beta T} \tag{4}$$

where R is the component respiration (soil ($\mu mol \cdot m^{-2} \cdot s^{-1}$), root ($\mu mol \cdot m^{-2} \cdot s^{-1}$), coarse woody debris ($\mu mol \cdot m^{-3} \cdot s^{-1}$), stem ($\mu mol \cdot m^{-3} \cdot s^{-1}$), or leaf ($\mu mol \cdot g^{-1} \cdot d^{-1}$)); T is the temperature of each component ($^\circ$C); and R_0 and β are fitted parameters. The temperature dependence of respiration is often described by the Q_{10} value, which is called the temperature sensitivity of respiration. The respiration parameter, Q_{10}, can be derived from $Q_{10} = \exp(10\beta)$. Estimated parameters were used to predict the soil respiration for every 0.5 h over 4 years based on continuous temperature measurements from the weather station.

2.2.7. Root Respiration

The trenching method was used to estimate the root respiration [33]. The trenched plot (20 m × 20 m) was established adjacent to the permanent plot at this site. We also randomly established twenty 50 cm × 50 cm subplots in the trenched plot in August 2009. Each subplot was prepared by making vertical cuts along the boundaries to 50 cm below the ground surface (approximately the bottom of the root zone) with a steel knife, severing all roots. The roots were removed, and fiberglass sheets were installed to prevent roots from entering. The trenches were backfilled with the same soil. The aboveground parts of all plants growing in the subplots were cut off, and new seedlings and re-growth from the roots were periodically clipped when necessary.

Twenty soil collars, each with a height of 10 cm and a diameter of 10 cm, were inserted into the soil in the subplots. The soil respiration in the trenched plot was measured using the same method for soil respiration.

We used the following equation to calculate the root respiration (R_R, $\mu mol \cdot m^{-2} \cdot s^{-1}$):

$$R_R = R_S - R_C \tag{5}$$

where R_S is the soil respiration in the permanent plot ($\mu mol \cdot m^{-2} \cdot s^{-1}$) and R_C is the soil respiration in the trenched plot ($\mu mol \cdot m^{-2} \cdot s^{-1}$).

2.2.8. Coarse Woody Debris Respiration

We used the standard method developed by the United States Department of Agriculture (USDA) Forest Service and the Long Term Ecological Research (LTER) programme to define woody debris as CWD, which was further categorized into logs, snags, and stumps [34]. The downed or leaning deadwood with a diameter at the widest point $\geq$10 cm and length $\geq$1 m were included in the group. The dead trees with a gradient (departure from vertical direction) $\leq$45° were considered as snags, while those with a gradient >45° were classified as logs. The vertical deadwood with a height $\leq$1 m was considered as stumps. Each piece of CWD was assigned to one of five decay classes on the basis of differences in internal and external tissue characteristics (Table S4) [35]. The numbers 1, 2, 3, 4, and 5 represent different decomposition stages, i.e., 1 represents the initial stage and 5 represents the final stage.

CWD respiration was measured for the five decay classes in the plot. Three pieces of CWD were sampled for CWD respiration in each decay class, and three fixed plates were mounted on each low decay class of CWD (sufficient sound wood was present) with silicon sealant at a random azimuth. A custom Plexiglas cuvette, 800 cm^3 in volume with an 80 cm^2 opening, was closely attached to the mounting plate just before each measurement. CWD respiration was measured three times in succession for each cuvette during each measurement and three times during the day at each cuvette. CWD temperature at 10 cm deep was measured adjacent to each cuvette with a portable temperature probe provided with the LI-6400. For the more advanced decay classes, these CWD samples were placed into the containers to measure. The measurements were made every 2 weeks from April to November in 2010–2013.

The measured CWD respiration rates per unit area were converted to rates per unit volume. We used the exponential function (Q_{10} function) to analyze the response of CWD respiration per unit of volume to CWD temperature (Equation (4)). Continuous CWD temperatures were calculated by the model, which simulated the relationship between CWD temperatures and 10 cm soil temperatures (Figure S5).

To upscale the chamber measurements of CWD respiration to the stand level, we calculated the volume of the five decay classes of CWD in the plot. Forest censuses were conducted in August of each year during 2009–2013 to determine the CWD volume. Each log or stump was considered as a cylinder; consequently, we used Smalian's formula to produce a volume estimate through the length and cross-sectional areas at the basal and distal ends of the cylinder [36]. For snags, we used the height and diameter in a species-specific wood volume equation, thus calculating the volume of each piece of the snag.

2.2.9. Stem Respiration

Fifty fixed plates were mounted on the trunks of 50 standard trees with silicon sealant at approximately 130 cm in height and a random azimuth. We used the same cuvette that was used to measure CWD respiration for measuring stem respiration; the cuvette was closely attached to the mounting plate just before each measurement. For the CWD respiration measurements, stem respiration rates were measured three times in succession for each cuvette during each measurement and three times during the day at each cuvette. The measurements were made every 2 weeks with an LI-6400 portable photosynthesis system (LI-COR, Inc., Lincoln, NE, USA) from April to November in 2010–2013. Stem temperature was measured with a portable temperature probe provided with the LI-6400 inserted into the sapwood near the cuvette of each sample tree. The sapwood thickness and wood mass density of each standard tree were measured from tree cores.

Measured stem respiration rates per unit area were converted to rates per unit of sapwood volume based on sapwood area and tree DBH, assuming a wedge-shape volume had contributed to the respiration rates. We used the exponential function (Q_{10} function) to analyze the response of stem respiration per unit of sapwood volume to stem temperature (Equation (4)). Continuous stem temperatures were calculated by the model, which simulated the relationship between stem temperatures and air temperatures (Figure S6).

To upscale the chamber measurements of stem respiration to the stand level, we estimated the total sapwood volume per unit of ground area in the plot. We assumed that branch respiration per volume had the same rate as stem (bole) respiration, similar to the assumptions made by Law et al. [37], Xu et al. [38], and Bolstad et al. [4].

After measuring the DBH and sapwood thickness, we estimated the sapwood volume of 30 sample trees to develop the regression model for sapwood volume:

$$\ln V_p = 0.90589 \ln (D^2H) - 10.31542 \tag{6}$$

where V_p is the sapwood volume including that from stems and branches (m^3), D is the DBH (m), H is the tree height (m), and the correlation coefficient is 0.9452. Equation (6) was used to estimate the sapwood volume of the whole stand and the average sapwood volume per ground area.

2.2.10. Leaf Respiration

Leaf respiration was measured from 30 leaves collected from 10 *L. principis-rupprechtii* trees, 30 leaves from shrubs, and 30 leaves from herbs from April to November in 2010–2013. Following the method of Bolstad et al. [4], branches from species from random heights and directions in the canopy were detached at night and immediately placed in a plastic bag with a moistened paper towel and transported in the dark to a nearby laboratory. Fully expanded leaves were detached just before measurement. All measurements were made within 3 h of branch harvest. Leaf respiration rates were measured from 5 to 25 °C with a controlled temperature LI-6400 portable photosynthesis system (LI-COR, Inc., Lincoln, NE, USA). Leaf area was measured with an AM-300 portable leaf area meter (ADC Bioscientific Limited, SG12, 9TA, Cambridge, UK). Leaves were oven dried at 65 °C and weighed. The measured leaf respiration rates per unit area were converted to rates per unit of dry biomass.

We used the exponential equation (Equation (4)) to fit the leaf respiration per unit of dry biomass as a function of leaf temperature for each species. We assumed in this study that the leaf temperature was the air temperature.

To upscale chamber measurements of leaf respiration to the stand level, we estimated the total leaf dry biomass per unit of ground area in the plot. The dry leaf biomass of shrubs and herbs was estimated by our study, while the dry leaf biomass of *L. principis-rupprechtii* was estimated by the regression model based on a previous study in this region [28].

2.2.11. Net Ecosystem Exchange

The annual net ecosystem exchange of CO_2 (NEE, $g \cdot C \cdot m^{-2} \cdot y^{-1}$) can be estimated using the following equation according to the measured annual rates of component respirations and net primary production (NPP):

$$NEE = R_S + R_{CWD} - R_R - NPP \tag{7}$$

where R_S is the soil respiration ($g \cdot C \cdot m^{-2} \cdot y^{-1}$), R_{CWD} is the CWD respiration ($g \cdot C \cdot m^{-2} \cdot y^{-1}$), Equation (5) was used to calculate the R_R ($g \cdot C \cdot m^{-2} \cdot y^{-1}$), and Equation (1) was used to calculate the NPP ($g \cdot C \cdot m^{-2} \cdot y^{-1}$).

2.3. Eddy Covariance Measurements

To compare with biometric measurements, fluxes of CO_2 were measured from a tower at 30 m above ground in the *center* of the stand. A three-dimensional sonic anemometer (CSAT-3, Campbell Scientific, Inc., Logan, UT) and an open-path infrared gas analyzer (LI-7500, LI-COR, Lincoln, NE, USA) mounted at a height of 25 m measured the three components of the wind velocity vector, sonic temperature, and the densities of water vapor and CO_2. These components were sampled at 10 Hz by a data logger (CR-5000, Campbell Scientific, Logan, UT, USA), which also calculated the 30 min covariance using Reynolds block averaging. Surface fluxes were later calculated off-line after performing a two-dimensional coordinate rotation and accounting for density fluctuations [39]. NEE data were screened for weak turbulence friction velocity at night. Although we found only a negligible trend of increasing NEE with u^*, we calculated an annual NEE using a u^* threshold of $0.15 \text{ m} \cdot \text{s}^{-1}$. To fill the gaps, a double-directional interpolation model of artificial neural networks (ANNs) was used [27]. Nighttime NEE was assumed to be a measurement of ecosystem respiration and was extrapolated to all times by using a temperature response function as described by Cook et al. [40] and Desai et al. [41].

2.4. Statistical Analyses

One-way ANOVAs were used to determine the effect of the 10 cm soil temperature on the soil respiration, of CWD temperature on the respiration of different CWD decay classes, of sapwood temperature on the stem respiration, and of air temperature on the leaf respiration. An exponential equation was used to simulate the relationship between respiration and temperature. The relationship between the CWD temperature and the 10 cm soil temperature was simulated using a regression model. A regression model was also used to simulate the relationship between sapwood temperature and air temperature. The sapwood volume was estimated by a regression model. Moreover, the eddy covariance technique and chamber-based estimates were simulated based on a linear model. All statistical analyses were conducted using the SAS 8.0 Statistical Package, with a *p*-value of 0.05 set as the limit for statistical significance. Origin 8.0 (OriginLab Corporation, Northampton, MA, USA) was used to draw the graphs.

3. Results

3.1. Environmental Factors

There was a clear seasonal pattern in air temperature and 10 cm soil temperature during 2010–2013 (Figure S7). The air temperature changed more dramatically, but the variation of the 10 cm soil temperature was consistent with the air temperature. The annual mean air temperature was 10.82 ± 9.66, 10.94 ± 9.78, 10.59 ± 9.60, and 10.89 ± 9.84 °C for 2010 to 2013, respectively. The annual mean 10 cm soil temperature was 11.12 ± 8.09, 11.22 ± 8.19, 10.93 ± 8.04, and 11.18 ± 8.24 °C for 2010 to 2013, respectively. Due to rain, the photosynthetic active radiation was relatively low from June to August for all years (Figure S8). The annual mean photosynthetically active radiation was 147.81 ± 92.09, 149.05 ± 103.63, 146.19 ± 94.49, and 144.54 ± 81.05 $\mu mol \cdot m^{-2} \cdot s^{-1}$ for 2010 to 2013, respectively.

3.2. Soil Respiration

There was a significant exponential relationship between soil respiration and the 10 cm soil temperature. The parameters in Equation (4) for soil respiration are summarized in Table 1. Equation (4) allows us to estimate the year-round soil respiration using the 10 cm soil temperature as an independent variable. There was an obvious seasonal pattern of soil respiration in 2010–2013 (Figure 1). Soil respiration includes soil heterotrophic respiration, root respiration, and litter respiration; root respiration accounted for $35\% \pm 7\%$ of the soil respiration on average in 2010–2013 (Figure 1). The total soil respiration was 710.37 ± 20.14, 721.45 ± 22.45, 692.77 ± 19.64, and 737.27 ± 24.27 $g \cdot C \cdot m^{-2} \cdot y^{-1}$ for 2010 to 2013, respectively. The lower soil respiration in 2012 was consistent with the soil temperature in 2012, which was lower than in other years.

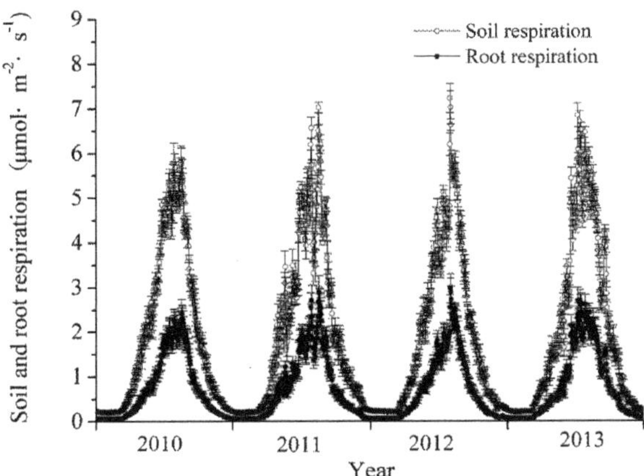

Figure 1. Daily mean soil and root respiration in the *L. principis-rupprechtii* forest during 2010–2013. Error bars are based on 0.5 h soil temperatures as experimental unit ($n = 48$).

Table 1. Parameters in the temperature response function (Equation (4)) for soil respiration (R_S, $\mu mol \cdot m^{-2} \cdot s^{-1}$), root respiration ($R_R$, $\mu mol \cdot m^{-2} \cdot s^{-1}$), coarse woody debris (CWD) respiration (R_{CWD}, $\mu mol \cdot m^{-3} \cdot s^{-1}$), stem respiration ($R_T$, $\mu mol \cdot m^{-3} \cdot s^{-1}$), and leaf respiration (R_L, $\mu mol \cdot g^{-1} \cdot d^{-1}$).

Respiration (R)		R_0	β	Q_{10}	Coefficient of Determination (R^2)	Samples (n)
R_S		0.22	0.14	4.19	0.76	5760
R_R		0.12	0.11	3.00	0.66	3840
R_{CWD}	1	1.03	0.12	3.44	0.72	576
	2	1.30	0.10	2.62	0.81	576
	3	3.25	0.07	2.06	0.73	576
	4	3.43	0.10	2.81	0.76	576
	5	3.03	0.10	2.59	0.82	576
R_T		1.42	0.09	2.51	0.72	9600
R_L	Tree	3.52	0.07	1.98	0.65	1920
	Shrub	4.81	0.06	1.85	0.68	1920
	Herb	2.75	0.07	1.92	0.65	1920

3.3. Coarse Woody Debris (CWD) Respiration

There was a strong exponential correlation between the respiration of different CWD decay classes and temperature (Table 1). To estimate continuous CWD respiration in the *L. principis-rupprechtii* forest during 2010–2013, we calculated continuous CWD temperature using the model that simulated the relationship between the CWD temperature and the 10 cm soil temperature. The seasonal variation of the CWD respiration per volume indicated that the CWD respiration peaked in July and then followed a decreasing trend with time (Figure 2). We found significant differences in the CWD respiration among the different decay classes in 2010–2013. The maximum respiration across the four years, with a mean of 15.33 ± 1.32 $\mu mol \cdot m^{-3} \cdot s^{-1}$, appeared in decay class 4, which was more than 3 times higher than the minimum in decay class 1 (4.65 ± 0.37 $\mu mol \cdot m^{-3} \cdot s^{-1}$). Over time, the volume of different CWD decay classes increased (Table 2). The respiration of different CWD decay classes per unit ground area also increased over time, except for a slight decrease in 2012 (Table 2), which was consistent with the temperature of different CWD decay classes.

Table 2. Volume ($m^3 \cdot ha^{-1}$) and the respiration of different coarse woody debris (CWD) decay classes per ground area ($g \cdot C \cdot m^{-2} \cdot y^{-1}$) during 2010–2013.

CWD Decay Classes		Year			
		2010	2011	2012	2013
1	Volume	16.83 (1.23)	17.90 (1.51)	18.17 (1.31)	19.21 (1.62)
	Respiration	2.85 (0.23)	3.06 (0.18)	3.02 (0.26)	3.31 (0.28)
2	Volume	14.64 (1.18)	14.90 (1.22)	15.89 (1.08)	16.41 (1.33)
	Respiration	3.03 (0.17)	3.12 (0.25)	3.21 (0.23)	3.49 (0.27)
3	Volume	8.95 (0.82)	9.28 (0.78)	9.40 (0.68)	9.82 (0.72)
	Respiration	2.84 (0.23)	2.97 (0.18)	2.94 (0.23)	3.16 (0.22)
4	Volume	6.39 (0.51)	6.63 (0.48)	6.71 (0.53)	7.03 (0.38)
	Respiration	3.69 (0.19)	3.88 (0.26)	3.80 (0.31)	4.16 (0.34)
5	Volume	5.11 (0.31)	5.30 (0.28)	5.38 (0.38)	5.65 (0.27)
	Respiration	2.16 (0.16)	2.27 (0.21)	2.23 (0.27)	2.44 (0.26)
Total	Volume	51.92 (3.86)	54.01 (4.51)	55.55 (4.39)	58.12 (4.88)
	Respiration	14.57 (1.58)	15.30 (1.35)	15.20 (1.61)	16.56 (1.48)

Note: standard error is provided in brackets.

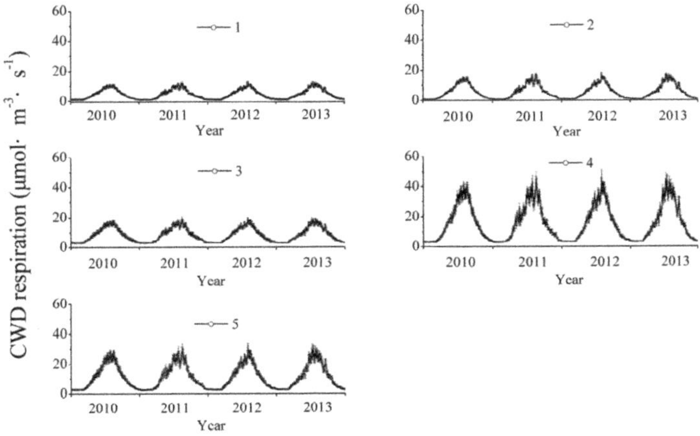

Figure 2. The daily mean respiration of different coarse woody debris (CWD) decay classes in the *L. principis-rupprechtii* forest during 2010–2013. Error bars are based on 0.5 h soil temperatures as experimental unit (n = 48).

3.4. Stem Respiration

We found an exponential relationship between stem respiration and sapwood temperature, and the parameters are summarized in Table 1. Because sapwood temperature was not recorded for the whole year, we developed a correlation between sapwood temperature (y) and air temperature (x) (y = 0.62385 + 0.88813x, R^2 = 0.98674, n = 108, p < 0.0001). Stem respiration varied seasonally, with the lowest rate in February and the highest in July (Figure 3). Sapwood volume was 189.21 ± 12.31, 193.52 ± 11.08, 196.28 ± 15.57, and 215.21 ± 18.11 $m^3 \cdot ha^{-1}$ during 2010 to 2013, respectively. We estimated the stem respiration per unit of ground area as 33.52 ± 2.86, 34.68 ± 3.18, 34.12 ± 2.17, and 38.82 ± 3.22 $g \cdot C \cdot m^{-2} \cdot y^{-1}$ from 2010 to 2013, respectively.

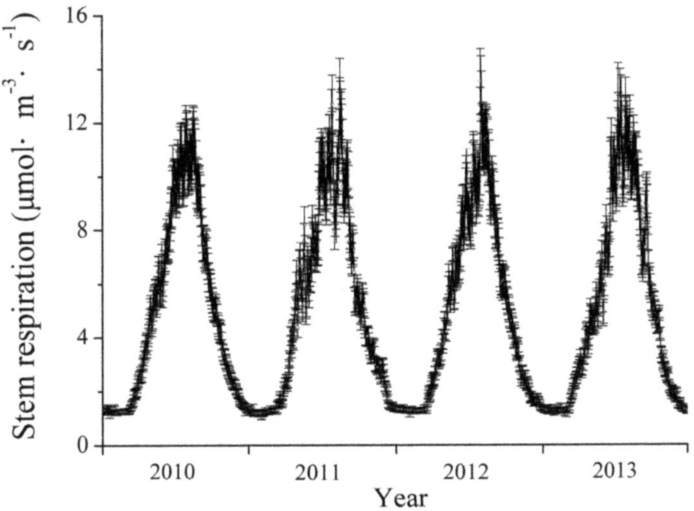

Figure 3. Daily mean stem respiration in the *L. principis-rupprechtii* forest during 2010–2013. Error bars are based on 0.5 h air temperatures as experimental unit (n = 48).

3.5. Leaf Respiration

The diurnal and seasonal variations in the leaf respiration rate related well with the corresponding variation in air temperature (Table 1). The minimum leaf respiration for trees, shrubs, and herbs all appeared in March, rose sharply until July, and then decreased until November (Figure 4). Shrub leaves per unit of biomass had slightly higher respiration than tree and herb leaves. Leaf biomass increased with time (Table 3), which was consistent with the LAI measured by LAI-2000. Cumulative leaf respiration per unit of ground area for trees, shrubs, and herbs indicated that tree leaves had higher respiration than shrub and herbaceous leaves (Table 3), corresponding mainly with the highest leaf biomass for trees. Over time, the leaf respiration per unit of ground area increased.

Table 3. Leaf dry biomass (t·ha^{-1}) and leaf respiration per ground area (g·C·m^{-2}·y^{-1}) for trees, shrubs, and herbs, and leaf area index (LAI) during 2010–2013.

Items		Year			
		2010	2011	2012	2013
Tree	Biomass	10.35 (0.86)	11.23 (0.78)	12.11 (1.14)	13.42 (1.21)
	Respiration	41.53 (3.89)	45.28 (4.15)	45.21 (3.71)	53.39 (4.82)
Shrub	Biomass	0.10 (0.04)	0.11 (0.03)	0.13 (0.02)	0.15 (0.01)
	Respiration	0.46 (0.03)	0.50 (0.04)	0.54 (0.06)	0.71 (0.08)
Herb	Biomass	0.42 (0.04)	0.45 (0.05)	0.48 (0.04)	0.51 (0.04)
	Respiration	1.28 (0.15)	1.37 (0.18)	1.40 (0.11)	1.56 (0.17)
All	Biomass	10.87 (1.04)	11.79 (0.82)	12.72 (1.14)	14.08 (1.31)
	Respiration	43.27 (3.92)	47.15 (4.27)	47.15 (4.11)	55.66 (4.52)
LAI		2.15 (0.56)	2.47 (0.62)	2.74 (0.46)	3.15 (0.48)

Note: standard error is provided in brackets.

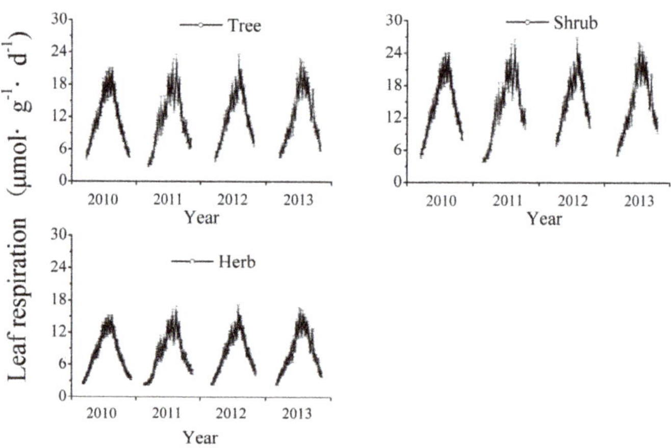

Figure 4. Daily mean leaf respiration for trees, shrubs, and herbs in the *L. principis-rupprechtii* forest during 2010–2013. Error bars are based on 0.5 h air temperatures as experimental unit (*n* = 48).

3.6. Ecosystem Respiration

The minimum ecosystem respiration appeared in January–March, rose sharply until July, and then decreased until winter (Figure 5). Ecosystem respiration ranged from 0.22 ± 0.02 to 7.86 ± 0.38 μmol·m^{-2}·s^{-1} in 2010–2013. The component respiration demonstrated a similar seasonal variation with ecosystem respiration. Soil was the strongest source of annual respiration for the ecosystem, which was approximately 50 times higher than the minimum respiration for CWD (Table 4). Soil, CWD, stem, leaf, and ecosystem respiration in 2013 was the highest among 2010–2013. Aboveground autotrophic respiration (stem + leaf respiration) comprised 10.26% of the total respiration, with leaf respiration slightly higher than stem respiration.

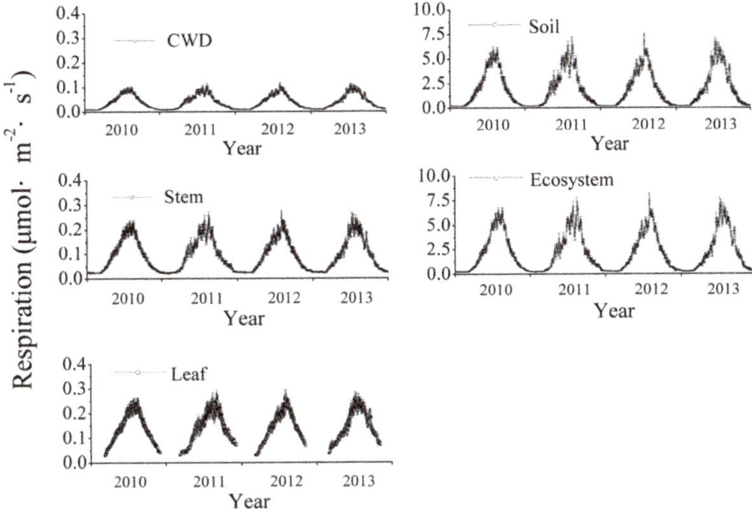

Figure 5. Daily mean soil, CWD, stem, leaf, and total ecosystem respiration in the *L. principis-rupprechtii* forest during 2010–2013. Error bars are based on 0.5 h soil temperatures and air temperatures as experimental unit, respectively ($n = 48$).

Table 4. Ecosystem respiration, component respiration ($g \cdot C \cdot m^{-2} \cdot y^{-1}$), and percentage (%) during 2010–2013 and the average over four years.

Year	Soil	CWD	Stem	Leaf	Ecosystem
2010	710.37 (20.14)	14.57 (1.58)	33.52 (2.86)	43.27 (3.92)	801.73 (58.67)
Percentage	88.60%	1.82%	4.18%	5.40%	100%
2011	721.45 (22.45)	15.30 (1.35)	34.68 (3.18)	47.15 (4.27)	818.58 (68.14)
Percentage	88.13%	1.87%	4.24%	5.76%	100%
2012	692.77 (19.64)	15.20 (1.61)	34.12 (2.17)	47.15 (4.11)	789.24 (51.55)
Percentage	87.78%	1.93%	4.32%	5.97%	100%
2013	737.27 (24.27)	16.56 (1.48)	38.82 (3.22)	55.66 (4.52)	848.31 (74.12)
Percentage	86.91%	1.95%	4.58%	6.56%	100%
Average	715.47 (28.48)	15.41 (1.72)	35.28 (4.78)	48.31 (5.24)	814.47 (64.22)
Percentage	87.85%	1.89%	4.33%	5.93%	100%

Note: standard error is provided in brackets.

3.7. Net Primary Productivity

The carbon content ratio was measured to calculate the carbon density, and the carbon content ratio was different in various components of vegetation in the *L. principis-rupprechtii* forest (Table S9). NPP was highest in 2013 (833.33 ± 80.14 $g \cdot C \cdot m^{-2} \cdot y^{-1}$), with an average of 817.16 ± 81.48 $g \cdot C \cdot m^{-2} \cdot y^{-1}$ in the *L. principis-rupprechtii* forest, of which ΔB_{living} and D_{total} accounted for 77.7%, and 22.3%, respectively (Table 5). NEE was the lowest in 2012 (-344.14 ± 22.32 $g \cdot C \cdot m^{-2} \cdot y^{-1}$), and the average in all years was -336.71 ± 25.15 $g \cdot C \cdot m^{-2} \cdot y^{-1}$, which revealed a carbon sink in the *L. principis-rupprechtii* forest.

Table 5. Annual net primary productivity (NPP) and net ecosystem exchange (NEE) during 2010–2013 ($g \cdot C \cdot m^{-2} \cdot y^{-1}$).

Items	2010	2011	2012	2013	Average
ΔB_{living}	656.48 (51.22)	623.35 (52.25)	605.88 (51.14)	654.15 (53.33)	634.97 (52.52)
D_{total}	155.04 (12.85)	190.82 (18.17)	203.76 (18.82)	179.18 (15.69)	182.19 (16.85)
NPP	811.52 (72.28)	814.17 (67.58)	809.64 (72.66)	833.33 (80.14)	817.16 (81.48)
NEE	−335.21 (20.52)	−329.93 (19.64)	−344.14 (22.32)	−337.55 (17.34)	−336.71 (25.15)

Note: standard error is provided in brackets.

3.8. Comparison between Biometric and Eddy Covariance Measurements

Eddy covariance measurements of NEE showed seasonal variation, with a maximum in November (2.71 ± 0.22 $g \cdot C \cdot m^{-2} \cdot d^{-1}$) and a minimum in August (-6.84 ± 0.56 $g \cdot C \cdot m^{-2} \cdot d^{-1}$) (Figure 6). Eddy covariance measurements of NEE from May to September were negative, which suggested a carbon sink during the growing season. Overall, eddy covariance measurements of NEE showed an average of -288.33 ± 25.26 $g \cdot C \cdot m^{-2} \cdot y^{-1}$ (-292.28 ± 18.17 $g \cdot C \cdot m^{-2} \cdot y^{-1}$, -281.02 ± 17.22 $g \cdot C \cdot m^{-2} \cdot y^{-1}$, -270.56 ± 20.21 $g \cdot C \cdot m^{-2} \cdot y^{-1}$, and -309.47 ± 24.37 $g \cdot C \cdot m^{-2} \cdot y^{-1}$ for 2010 to 2013, respectively), which indicated a sink at this site. These numbers are close to the results from the biometric measurements but average 14% higher. Based on eddy covariance measurements, annual ecosystem respiration was estimated as 780.37 ± 82.18 $g \cdot C \cdot m^{-2} \cdot y^{-1}$ (759.12 ± 62.51 $g \cdot C \cdot m^{-2} \cdot y^{-1}$, 766.38 ± 71.17 $g \cdot C \cdot m^{-2} \cdot y^{-1}$, 781.49 ± 64.82 $g \cdot C \cdot m^{-2} \cdot y^{-1}$, and 814.48 ± 75.18 $g \cdot C \cdot m^{-2} \cdot y^{-1}$ for 2010 to 2013, respectively). Daily mean ecosystem respiration based on chamber measurements and eddy covariance measurements are plotted in Figure 7. The curve shows a good correlation between these two measurement methods, with $R^2 = 0.93$.

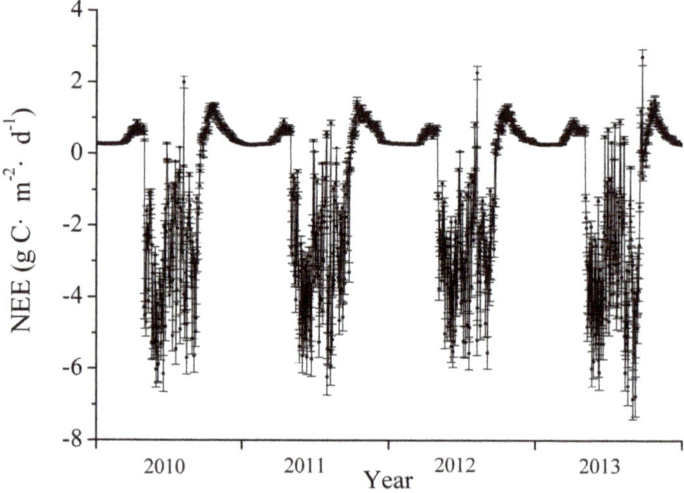

Figure 6. Eddy covariance measurements of NEE in the *L. principis-rupprechtii* forest. Data are daily mean NEE during 2010–2013. Error bars are based on 0.5 h eddy covariance measurements of NEE as experimental unit ($n = 48$).

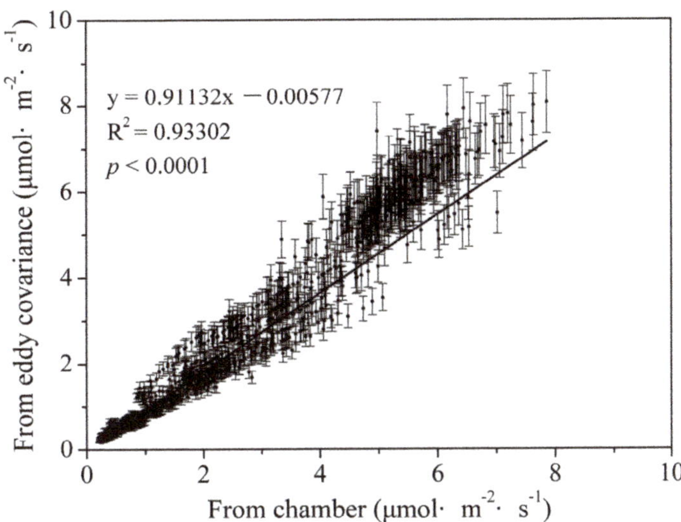

Figure 7. Comparison between chamber-based and eddy covariance measurements of ecosystem respiration. Data are daily mean respiration during 2010–2013. The solid line indicates a linear fit.

4. Discussion

The seasonal variation of component respiration was driven by temperature, and temperature was strongly exponentially correlated with component respiration. This conclusion has been reported in several studies [16,26]. According to our analysis, the ecosystem respiration based on eddy covariance measurements was also exponentially related to the air temperature, explaining 94% of the variability in the ecosystem respiration (Figure S10).

Biometric-based flux measurements combined with spatial and temporal upscaling allowed us to estimate the respiration of each component of the ecosystem. The CO_2 flux of soil from our site was higher than that from a mature temperate mixed forest in the Changbai Mountains (0.49–4.12 $\mu mol \cdot m^{-2} \cdot s^{-1}$), which may be due to the higher temperature at our site (Annual average temperature is 3.6 °C in the Changbai Mountains) [26]. Soil respiration is influenced by many factors, including abiotic factors, such as the soil temperature, soil moisture, rainfall, and soil C/N, and biotic factors, such as vegetation cover, litter thickness on the ground, fine root mass, soil organic matter, soil characteristics, leaf area index, and human activities. A number of studies concluded that soil moisture is an important factor influencing soil respiration [42–44]. However, a comparison of recent findings in this region showed that soil moisture had little impact on soil respiration relative to in situ temperature. An adequate soil moisture level may explain the non-significant influence of soil moisture on soil respiration [26]. The Q_{10} value was the highest in soil respiration, with a value of 4.19, which revealed that there was a stronger sensitivity of soil respiration to temperature in different component respirations, which was different from other ecosystems. For example, Raich and Schlesinger reviewed global soil respiration values and found a median Q_{10} value of 2.4 [45]. In a temperate mixed-hardwood forest, the Q_{10} value ranged from 3.4 to 5.6 [46]. The Q_{10} value ranged from 1.93 to 4.80 in different vegetation communities in the Belgian Campine region [47]. Moreover, Cao et al. reported Q_{10} values of 2.75 and 3.22 in an alpine meadow on the Tibetan Plateau of China [48]. Wu et al. reported a value of 4.07 in the Changbai Mountains of China [49]. On the eastern part of the Loess Plateau of China, the Q_{10} value ranged from 2.37 to 5.53 [50]. The various Q_{10} values reported in different ecosystems may be due to both methodology and differences in biotic and abiotic factors that affect soil respiration differently [50].

Whole-ecosystem respiration is dominated by soil respiration in forests [4]. Soil respiration has been found to account for 30–90% of total ecosystem respiration in temperate forests [13], and our result was close to the upper limit, at 87.85%. The percentage of soil respiration was higher, possibly due to the lower precipitation (600 mm) at our site, but there were lower nutrients in the soil (sandy loam) at a Michigan site and in a ponderosa pine forest in Oregon (76%) [37]. Total soil respiration is always regarded as a critical component of the carbon balance of an ecosystem. The estimated annual soil respiration was 715.47 ± 28.48 g·C·m^{-2}·y^{-1}, which was similar to that reported by Raich and Schlesinger for temperate coniferous forests, at 681 ± 95 g·C·m^{-2}·y^{-1} [45]. Other ecosystems reported various values for soil respiration, including 600–700 g·C·m^{-2}·y^{-1} in beech forests in France [51], 485 g·C·m^{-2}·y^{-1} in hardwood forests in the USA [52], 581 g·C·m^{-2}·y^{-1} in a warm-temperate mixed forest in Japan [53], 963.98 g·C·m^{-2}·y^{-1} in a Korean pine and broadleaf mixed forest in China [54], and 454 g·C·m^{-2}·y^{-1} in a tropical seasonal rainforest in China [55]. These different results demonstrate that soil respiration is closely related to vegetation and climate.

In our study, leaf respiration was second in magnitude among the respiration fractions. This result is consistent with that of previous studies [5,26]. However, the leaf respiration at our site was lower than that in an old-growth hardwood forest (114.85 g·C·m^{-2}·y^{-1}) and a hemlock forest (72.20 g·C·m^{-2}·y^{-1}) in the USA [16]. Moreover, Law et al. compared leaf respiration in young (124 g·C·m^{-2}·y^{-1}) and old (136 g·C·m^{-2}·y^{-1}) ponderosa pine forests of the USA [56], which were all higher than ours. In the present study, this lower leaf respiration may have been caused by (1) the lower annual mean LAI (4.1 in the hardwood forest and 3.8 in the hemlock forest), (2) reduced air temperature and photosynthetic active radiation [57], or (3) decreased foliage N and chlorophyll concentrations [37]. In addition, both the canopy position and foliage age had significant effects on leaf respiration [12].

Because of difficulties in measurement, little attention was paid to stem respiration in the past. However, with increasing atmospheric CO_2 concentrations and as an important part of the annual carbon balance of forest ecosystems, studies of stem respiration have generated active debate. Stem respiration was strongly correlated with temperature, and the Q_{10} value at our site was similar to that of other ecosystems. For example, in a *Pinus koraiensis* forest in China, the Q_{10} value ranged from 2.56 to 3.32 [58]. In a temperate mixed forest in China, the Q_{10} value ranged from 1.86 to 2.41 [26]. Ryan et al. reported that the Q_{10} value ranged from 1.3 to 1.9 in four conifer forests in the USA [59]. According to our study, stem respiration was slightly lower than leaf respiration, which was consistent with others [5,26,37,38]. However, the stem respiration at our site was much lower than those in an old-growth hardwood forest (130.5 g·C·m^{-2}·y^{-1}) and a hemlock forest (206.5 g·C·m^{-2}·y^{-1}) in the USA [16], which may be due to lower sapwood volume (296.3 m^3·ha^{-1} in the hardwood forest and 448.5 m^3·ha^{-1} in the hemlock forest). In addition, stem respiration is influenced by numerous factors, including meteorological factors (e.g., stem temperature, CO_2 concentration, and humidity) and biological factors (tree species, tree age, diameter at breast height, sapwood size, and nitrogen content in the stem). Latitude, altitude, and topographic factors indirectly influence respiration rates through meteorological or biological factors [60]. Moreover, scaling up based on measurements of the stem respiration rate and sapwood volume may underestimate total stem respiration rates because younger woody tissues have higher respiration rates [38]. Future studies are required to develop a more accurate scaling-up scheme.

In general, CWD has high ecological relevance, contributes significantly to crucial ecological processes in forest ecosystems, and plays an essential role in carbon pools [61]. Furthermore, CWD respiration as a component of ecosystem respiration is essential to determining a forest's carbon budget [62]. However, few studies of CWD respiration have been conducted in most forest types [63]. In particular, quantification of the contribution of CWD respiration to total ecosystem respiration in the Qinling Mountains has not been previously conducted. Our study showed that the seasonality of CWD respiration was mainly driven by the CWD temperature and showed an overall bell-shaped curve for all five decay classes. This conclusion is corroborated by previous studies [64,65]. The Q_{10} ranged from 1.7 to 4.1 for different decay classes in various ecosystems [62,65], and our result fell within this range.

Moreover, Wu et al. reported that the Q_{10} value ranged from 2.41 to 2.95 in the Changbai Mountains of China [66]. In a montane moist evergreen broadleaf forest of China, the Q_{10} value ranged from 1.73 to 2.08 [67]. Furthermore, the Q_{10} value was significantly affected by the temperature ranges. For example, the Q_{10} value was 4.1 for 5–20 °C and 1.7 for 20–42 °C in a boreal black spruce forest in Canada [65].

CWD respiration is a complex process that depends on many factors, including tree species, temperature, moisture, substrate quality, diameter class, and decomposer type [68,69]. According to our study, the annual CWD respiration was much lower than that in an old-growth Amazonian forest (171.8 $g \cdot C \cdot m^{-2} \cdot y^{-1}$) [70], which may be mainly caused by the reduced CWD biomass at our site. Meanwhile, the annual CWD respiration was slightly lower than that in an old-growth hardwood forest (43 $g \cdot C \cdot m^{-2} \cdot y^{-1}$) and a hemlock forest (29 $g \cdot C \cdot m^{-2} \cdot y^{-1}$) in the USA [16]. The reason for higher annual CWD respiration in hardwood and hemlock forests may be the surface area of CWD for upscaling [16]. In our study, we quantified the CWD into five decay classes. The CWD respiration rates measured per unit volume might underestimate the CWD respiration rates, whereas CWD respiration rates measured per unit surface area might overestimate the CWD respiration rates. A large log may have large volume but smaller surface area than many small logs. As such, the conversion of CWD respiration rates measured per unit area to rates per unit volume might be more appropriate to CWD respiration in this forest. Moreover, to upscale chamber measurements of CWD respiration to the stand level, the volume was more convenient and accurate than the surface area. Measuring the CWD respiration accurately is important for estimating forest ecosystem respiration. Thus, further research should be done by using the surface area or volume of CWD for upscaling. In addition, future studies are necessary to measure CWD respiration in different diameter classes.

Daily mean ecosystem respiration measured from eddy covariance and chamber methods was shown in Figure 7. The eddy covariance (y) and chamber (x) methods were in better agreement after adjusting the CO_2 flux in this forest using the equation $y = 0.91132x - 0.00577$ (Figure 7). The comparisons between chamber and eddy covariance ecosystem respiration measurements were more consistent than in other studies [71,72]. Our result showed that the daily mean ecosystem respiration upscaled from chamber measurements agreed well with eddy covariance measurements, with $R^2 = 0.93$, which was similar to that of an old-growth forest in the Great Lakes region of the USA [16], with $R^2 = 0.96$, although some studies suggested that the result from chamber measurements was higher than that from eddy covariance measurements [37,73].

An annual average ecosystem respiration of 780.37 ± 82.18 $g \cdot C \cdot m^{-2} \cdot y^{-1}$ was calculated from eddy covariance measurements. Based on chamber measurements, the corresponding value was 814.47 ± 64.22 $g \cdot C \cdot m^{-2} \cdot y^{-1}$. Our result is very close to that in a boreal black spruce forest in Canada, which had a range of 790–890 $g \cdot C \cdot m^{-2} \cdot y^{-1}$ [74]. However, ecosystem respiration from our site was much lower than that from old-growth Amazon tropical forests, with an estimate of 2337.6 $g \cdot C \cdot m^{-2} \cdot y^{-1}$ [75]. Higher temperatures, longer growing seasons, and higher photosynthesis and growth rates in tropical forests may explain the higher respiration than that from our temperate coniferous forests.

Our study found that the annual average ecosystem respiration measured by the chamber method was 5% higher than that measured by the eddy covariance. The error sources for explaining this discrepancy are complicated. Firstly, chambers may disturb the environment and alter CO_2 concentrations, as well as pressure gradients, turbulent fluctuations, and air flow. Thus, they may interfere with the production and transport of CO_2 [26]. Closed chambers completely cover the ecosystem during the measurement process and thereby alter the natural long-wave radiation balance to almost zero. This causes reduced surface cooling, weak development of stable stratification, and, finally, higher respiration than those obtained via eddy covariance measurements [76]. Secondly, our annual average ecosystem respiration estimates based on chamber measurements may have uncertainties. One possible source of uncertainty is in the method of extrapolating a temperature-respiration relationship to the winter. Under a 95% confidence interval, the uncertainties in predicting daily mean soil respiration, CWD respiration, stem respiration, and leaf respiration averaged 10%, 8%, 12%, and 14%, respectively, of the predicted values. The uncertainties may also derive from the estimation of CWD and stem volumes, the dry leaf biomass, and the chamber upscaling processes. Thirdly, nocturnal eddy covariance measurements with low friction velocity, extrapolation of nighttime respiration to daytime respiration may underestimate the ecosystem respiration by the eddy covariance method. Annual NEE estimates were as high as -336.71 ± 25.15 g·C·m^{-2}·y^{-1} from biometric measurements but -288.33 ± 25.26 g·C·m^{-2}·y^{-1} from eddy covariance measurements. Our results revealed that this site was a carbon sink, which was consistent with the findings of Zhou et al. [11]; they found that a carbon sink in a *L. principis-rupprechtii* forest in China was -271 g·C·m^{-2}·y^{-1}. Meanwhile, our estimations of NEE fall well within the range reported between -70 and -740 g·C·m^{-2}·y^{-1} for temperate forests [15,77].

Our study found that the NEE measured by the eddy covariance method was 14% higher than that measured by the biometric, which is consistent with Wang et al. [26]. However, our study showed that the discrepancy was lower than that of Wang et al. (22.5%), and the mean difference (48.38 ± 9.67 g·C·m^{-2}·y^{-1}) was also lower than that of temperate forests (100 g·C·m^{-2}·y^{-1}) [78], which indicated that these two methods had good agreement in measuring the NEE in this forest. However, there was still a discrepancy between the biometric and eddy covariance measurements, which may be due to the complicated error sources.

The error in the result from both biometric and eddy covariance was associated with sampling methods of flux measurement [30]. Ehman et al. concluded that the largest error source in the biometric method may be attributable to sampling (that is, inter-plot variability) [79]. All natural ecosystems are heterogeneous at some scale, and so the question arises whether estimation of fluxes based on a limited number of samples represents the average of the total ecosystem. In our study, the sampling error included estimating the NPP for sampling the trees, shrubs, herbs, fine roots, and litterfall. Additionally, we used soil collars to measure R_S and R_R and sampled the CWD to measure R_{CWD}, which all may produce the sampling error. In addition, Ohtsuka et al. considered that the topography may introduce sampling errors when estimating the NEE based on the eddy covariance method [30].

Another error was more likely to be related to estimation of NPP. Although Ohtsuka et al. deemed annual forest census within a large permanent plot and adequate number of litter traps for detritus production to be the most suitable method for measuring NPP [30], there were still some uncertainties. One possible uncertainty was derived from the biomass regression model. Under a 95% confidence interval, the uncertainties in predicting tree and shrub biomass averaged 8% and 14%, respectively, of the predicted values. Clark et al. considered that using off-site allometric equations could alter NPP by as much as 20% [80]. Also, these allometric relationships do not take into account seasonal changes in wood C concentration that may occur. Our study showed that the carbon content ratio values were diverse among varying plant species and different organs in the same plant (Table S9). However, many researchers used 50% as the carbon content ratio [81,82], which would overestimate the NPP by 7.4% in this forest. D_{total} is an important component of NPP; a study by Curtis et al. revealed that the detritus made up nearly two-thirds of the annual C production [77]. However, in our study, we found the D_{total} accounted for less than a quarter of NPP. CWD as a part of D_{total} can be difficult to measure precisely due to the heterogeneous distribution of forest floor detritus [79,80,83], which was the largest source of uncertainty in the estimation of NPP. However, we have adequately quantified annual variability in CWD production according to the annual forest census. Moreover, underestimation could have been caused by minor NPP components typically neglected (for example, root exudation, net accumulation of non-structural carbohydrates, herbivory consumption, production of volatile organic compounds [78,80]), which in general contribute between 1% and 4% of NPP [84]. In addition, Ohtsuka et al. considered that the estimation of R_{CWD} represented a critical source of potential error [30]; few studies have measured the R_{CWD}. Ehman et al. reported the R_{CWD} was 7% of R_S [79], but this proportion was only 2% in our study. We quantified the CWD into five decay classes to obtain more accurate measurements of R_{CWD}. However, the respirations of CWD at different diameter classes and fine woody debris (1 cm $\leq$ diameter < 10 cm) were not measured, which may result in a lower observed R_{CWD} at this site.

The permanent plot was protected by an enclosure, and this site was level (a mean slope of $5°$), and the overstory and understory of the forest were homogeneous, which are ideal conditions for measuring the NEE using the eddy covariance method in this forest. However, there were still error sources for measuring the NEE based on eddy covariance measurements. In our study, these eddy covariance error sources mainly include a double-directional interpolation model of artificial neural networks (ANNs) to fill the gaps, nocturnal low friction velocity, and the extrapolation of nighttime respiration from the eddy covariance measurements to daytime respiration.

5. Conclusions

This paper presents the first set of results (2010–2013) from a long-term project measuring forest-atmosphere CO_2 exchanges using eddy covariance and biometric methods simultaneously in a *L. principis-rupprechtii* forest. In our study, temperature was the primary controlling factor for respiration in this forest. Exponential functions explained most of the observed temporal variations in respiration in response to temperature. Based on the chamber upscale processes, we obtained the cumulative annual ecosystem respiration (814.47 ± 64.22 g·C·m^{-2}·y^{-1}), but this number was 5% higher than that from the eddy covariance measurements. We considered this small discrepancy to have arisen mainly because the respiration of CWD at different diameter classes and fine woody debris (1 cm $\leq$ diameter < 10 cm) was not measured due to the possible uncertainty in the method of extrapolating a temperature-respiration relationship to the winter and due to the uncertainties in the estimation of CWD and stem volumes, dry leaf biomass, and the chamber upscale processes.

Using measurements of the NPP (817.17 ± 81.48 g·C·m^{-2}·y^{-1}) combined with the heterotrophic respiration based on the chamber method (480.45 ± 52.24 g·C·m^{-2}·y^{-1}), we obtained the NEE (-336.71 ± 25.15 g·C·m^{-2}·y^{-1}) in this forest. This result was close to that from eddy covariance measurements (-288.33 ± 25.26 g·C·m^{-2}·y^{-1}), which indicated that these two methods had good agreement in measuring the NEE in this forest. In our study, we calibrated the instruments periodically, repeatedly measured a variety of respiration components for 4 years, reduced the disturbance of chamber measurements, quantified the CWD into five decay classes, and measured the carbon content ratio of various components, all of which can enhance the consistency of the eddy covariance and the biometric fluxes. However, there was still a 14% discrepancy between the biometric and eddy covariance measurements, which may be due to the complicated error sources. Thus, more detailed experiments and related theoretical studies are needed in the future.

Supplementary Materials: The following are available online at http://www.mdpi.com/1999-4907/9/2/67/s1, Figure S1: Some photos of *L. principis-rupprechtii* forest and flux tower in this region, Table S1: The regression model of biomass, volume, and height of *L. principis-rupprechtii*, Table S2: The regression model of shrub biomass in the *L. principis-rupprechtii* forest, Table S3: CWD characteristics of different decay classes in forest system, Figure S2: The relationship between temperature of different CWD decay classes and 10 cm soil temperature, Figure S3: The relationship between air temperature and stem temperature, Figure S4: Daily mean air temperature and 10 cm soil temperature in the *L. principis-rupprechtii* forest during 2010–2013, Figure S5: Daily mean photosynthetically active radiation in the *L. principis-rupprechtii* forest during 2010–2013, Table S4: Average carbon content ratio of litterfall and various organs of trees, shrubs, and herbs in the *L. principis-rupprechtii* forest (%), Figure S6: The relationship between ecosystem respiration based on eddy covariance measurement and air temperature. Data shown are daily means during 2010–2013.

Acknowledgments: We are grateful to the Qinling National Forest Ecosystem Research Station for providing some data and the experimental equipment. This research was funded by the project "Technical management system for increasing the capacity of carbon sink and water regulation of mountain forests in the Qinling Mountains" (201004036) of the State Forestry Administration of China.

Author Contributions: J.Y. and S.Z. conceived and designed the experiments; J.Y., Z.H., and J.P. performed the experiments; J.Y. analyzed the data; L.H. contributed reagents/materials/analysis tools; J.Y. and S.J. wrote the paper.

Conflicts of Interest: The authors declare no conflict of interest.

References

1. Post, W.M.; Peng, T.H.; Emanuel, W.R.; King, A.W.; Dale, V.H.; Angelis, D.L. The global carbon cycle. *Am. Sci.* **1990**, *78*, 310–326.

2. Conway, T.J.; Tans, P.P.; Waterman, L.S.; Thoning, K.W.; Kitzis, D.R.; Masarie, K.A.; Ni, Z. Evidence for interannual variability of the carbon cycle from the National Oceanic and Atmospheric Administration/Climate Monitoring and Diagnostics Laboratory Global Air Sampling Network. *J. Geophys. Res.* **1994**, *99*, 22831–22855. [CrossRef]

3. Houghton, R.; Hackler, J.; Lawrence, K. The US carbon budget: Contributions from land-use change. *Science* **1999**, *285*, 574–578. [CrossRef] [PubMed]

4. Bolstad, P.; Davis, K.; Martin, J.; Cook, B.; Wang, W. Component and whole-system respiration fluxes in northern deciduous forests. *Tree Physiol.* **2004**, *24*, 493–504. [CrossRef] [PubMed]

5. Guan, D.X.; Wu, J.B.; Zhao, X.S.; Han, S.J.; Yu, G.R.; Sun, X.M.; Jin, C.J. CO_2 fluxes over an old, temperate mixed forest in northeastern China. *Agric. For. Meteorol.* **2006**, *137*, 138–149. [CrossRef]

6. Kang, Y.X.; Chen, Y.P. Woody plant flora of Huoditang forest region. *J. Northwest For. Coll.* **1996**, *11*, 1–10. (In Chinese)

7. Schröter, D.; Cramer, W.; Leemans, R.; Prentice, I.C.; Araújo, M.B.; Arnell, N.W.; Bondeau, A.; Bugmann, H.; Carter, T.R.; Gracia, C.A.; et al. Ecosystem service supply and vulnerability to global change in Europe. *Science* **2005**, *310*, 1333–1337. [CrossRef] [PubMed]

8. Etzold, S.; Ruehr, N.K.; Zweifel, R.; Dobbertin, M.; Zingg, A.; Pluess, P.; Häsler, R.; Eugster, W.; Buchmann, N. The carbon balance of two contrasting mountain forest ecosystems in Switzerland: Similar annual trends, but seasonal differences. *Ecosystems* **2011**, *14*, 1289–1309. [CrossRef]

9. Leng, W.; He, H.S.; Bu, R.; Dai, L.; Hu, Y.; Wang, X. Predicting the distributions of suitable habitat for three larch species under climate warming in Northeastern China. *For. Ecol. Manag.* **2008**, *254*, 420–428. [CrossRef]

10. Lei, R.D.; Dang, K.L.; Zhang, S.X.; Tan, F.L. Effect of a *Larix principis-rupprechtii* forest plantation on soil in middle zone of south-facing slope of the Qinling mountains. *Sci. Silvae Sin.* **1997**, *33*, 463–470. (In Chinese)

11. Zhou, Y.R.; Yu, Z.L.; Zhao, S.D. Carbon storage and budget of major Chinese forest types. *Chin. J. Plant Ecol.* **2000**, *24*, 518–522. (In Chinese)

12. Brooks, J.R.; Hinckley, T.M.; Ford, E.D.; Sprugel, D.G. Foliage dark respiration in Abies amabilis (Dougl.) Forbes: Variation within the canopy. *Tree Physiol.* **1991**, *9*, 325–338. [CrossRef] [PubMed]

13. Valentini, R.; Matteucci, G.; Dolman, A.J.; Schulze, E.D.; Rebmann, C.; Moors, E.J.; Granier, A.; Gross, P.; Jensen, N.O.; Pilegaard, K.; et al. Respiration as the main determinant of carbon balance in European forests. *Nature* **2000**, *404*, 861–865. [CrossRef] [PubMed]

14. Malhi, Y.; Baldocchi, D.D.; Jarvis, P.G. The carbon balance of tropical, temperate and boreal forests. *Plant Cell Environ.* **1999**, *22*, 715–740. [CrossRef]

15. Baldocchi, D.D.; Wilson, K.B. Modeling CO_2 and water vapor exchange of a temperate broadleaved forest across hourly to decadal time scales. *Ecol. Model.* **2001**, *142*, 155–184. [CrossRef]

16. Tang, J.; Bolstad, P.V.; Desai, A.R.; Martin, J.G.; Cook, B.D.; Davis, K.J.; Carey, E.V. Ecosystem respiration and its components in an old-growth forest in the Great Lakes region of the United States. *Agric. For. Meteorol.* **2008**, *148*, 171–185. [CrossRef]

17. Khomik, M.; Arain, M.A.; Brodeur, J.J.; Peichl, M.; Restrepo Coupé, N.; Mclaren, J.D. Relative contributions of soil, foliar, and woody tissue respiration to total ecosystem respiration in four pine forests of different ages. *J. Geophys. Res. Biogeosci.* **2015**, *115*, 4209–4224. [CrossRef]

18. Grace, J.; Lloyd, J.; Mcintyre, J.; Miranda, A.C.; Meir, P.; Miranda, H.S.; Nobre, C.; Moncrieff, J.; Massheder, J.; Malhi, Y. Carbon dioxide uptake by an undisturbed tropical rain forest in southwest Amazonia, 1992 to 1993. *Science* **1995**, *270*, 778–780. [CrossRef]

19. Black, T.A.; Den Hartog, G.; Neumann, H.H.; Blanken, P.D.; Yang, P.C.; Russell, C.; Nesic, Z.; Lee, X.; Chen, S.G.; Staebler, R.; et al. Annual cycles of water vapour and carbon dioxide fluxes in and above a boreal aspen forest. *Glob. Chang. Biol.* **1996**, *2*, 219–229. [CrossRef]

20. Luyssaert, S.; Reichstein, M.; Schulze, E.D.; Janssens, I.A.; Law, B.E.; Papale, D.; Dragoni, D.; Goulden, M.L.; Granier, A.; Kutsch, W.L. Toward a consistency cross-check of eddy covariance flux-based and biometric estimates of ecosystem carbon balance. *Glob. Biogeochem. Cycles* **2009**, *23*, 1159–1171. [CrossRef]

21. Speckman, H.N.; Frank, J.M.; Bradford, J.B.; Miles, B.L.; Massman, W.J.; Parton, W.J.; Ryan, M.G. Forest ecosystem respiration estimated from eddy covariance and chamber measurements under high turbulence and substantial tree mortality from bark beetles. *Glob. Chang. Biol.* **2015**, *21*, 708–721. [CrossRef] [PubMed]

22. Goulden, M.L.; Munger, J.W.; Fan, S.M.; Daube, B.C.; Wofsy, S.C. Exchange of carbon dioxide by a deciduous forest: Response to interannual climate variability. *Science* **1996**, *271*, 1576–1578. [CrossRef]

23. Billesbach, D. Estimating uncertainties in individual eddy covariance flux measurements: A comparison of methods and a proposed new method. *Agric. For. Meteorol.* **2011**, *151*, 394–405. [CrossRef]

24. Davidson, E.; Savage, K.; Verchot, L.; Navarro, R. Minimizing artifacts and biases in chamber-based measurements of soil respiration. *Agric. For. Meteorol.* **2002**, *113*, 21–37. [CrossRef]

25. Myklebust, M.C.; Hipps, L.E.; Ryel, R.J. Comparison of eddy covariance, chamber, and gradient methods of measuring soil CO_2 efflux in an annual semi-arid grass, Bromus tectorum. *Agric. For. Meteorol.* **2008**, *148*, 1894–1907. [CrossRef]

26. Wang, M.; Guan, D.X.; Han, S.J.; Wu, J.L. Comparison of eddy covariance and chamber-based methods for measuring CO_2 flux in a temperate mixed forest. *Tree Physiol.* **2010**, *30*, 149–163. [CrossRef] [PubMed]

27. Dou, Z.Y.; Liu, J.J. Application of artificial neural networks to interpolation and extrapolation of flux data. *J. Northwest For. Univ.* **2009**, *24*, 58–62. (In Chinese)

28. Chen, C.G.; Peng, H. Standing crops and productivity of the major forest-types at the Huoditang forest region of the Qinling Mountains. *J. Northwest For. Coll.* **1996**, *11*, 92–102. (In Chinese, with English Abstract)

29. Rieger, I.; Lang, F.; Kleinschmit, B.; Kowarik, I.; Cierjacks, A. Fine root and aboveground carbon stocks in riparian forests: The roles of diking and environmental gradients. *Plant Soil* **2013**, *370*, 497–509. [CrossRef]

30. Ohtsuka, T.; Mo, W.; Satomura, T.; Inatomi, M.; Koizumi, H. Biometric based carbon flux measurements and net ecosystem production (NEP) in a temperate deciduous broad-leaved forest beneath a flux tower. *Ecosystems* **2007**, *10*, 324–334. [CrossRef]

31. Stenberg, P.; Linder, S.; Smolander, H.; Flower-Ellis, J. Performance of the LAI-2000 plant canopy analyzer in estimating leaf area index of some Scots pine stands. *Tree Physiol.* **1994**, *14*, 981–995. [CrossRef] [PubMed]

32. Cutini, A.; Matteucci, G.; Mugnozza, G.S. Estimation of leaf area index with the Li-Cor LAI 2000 in deciduous forests. *For. Ecol. Manag.* **1998**, *105*, 55–65. [CrossRef]

33. Munir, T.; Khadka, B.; Xu, B.; Strack, M. Partitioning forest-floor respiration into source based emissions in a boreal forested bog: Responses to experimental drought. *Forests* **2017**, *8*. [CrossRef]

34. Ringvall, A.; Ståhl, G. Field aspects of line intersect sampling for assessing coarse woody debris. *For. Ecol. Manag.* **1999**, *119*, 163–170. [CrossRef]

35. Yan, E.R.; Wang, X.H.; Huang, J.J.; Zeng, F.R.; Gong, L. Long-lasting legacy of forest succession and forest management, characteristics of coarse woody debris in an evergreen broad-leaved forest of eastern China. *For. Ecol. Manag.* **2007**, *252*, 98–107. [CrossRef]

36. Wenger, K.F. *Forestry Handbook*; John Wiley & Sons: New York, NY, USA, 1984.

37. Law, B.E.; Ryan, M.G.; Anthoni, P.M. Seasonal and annual respiration of a ponderosa pine ecosystem. *Glob. Chang. Biol.* **1999**, *5*, 169–182. [CrossRef]

38. Xu, M.; Debiase, T.A.; Qi, Y.; Goldstein, A.; Liu, Z. Ecosystem respiration in a young ponderosa pine plantation in the Sierra Nevada Mountains, California. *Tree Physiol.* **2001**, *21*, 309–318. [CrossRef] [PubMed]

39. Webb, E.K.; Pearman, G.I.; Leuning, R. Correction of flux measurements for density effects due to heat and water vapour transfer. *Q. J. R. Meteorol. Soc.* **1980**, *106*, 85–100. [CrossRef]

40. Cook, B.D.; Davis, K.J.; Wang, W.G.; Desai, A.; Berger, B.W.; Teclaw, R.M.; Martin, J.G.; Bolstad, P.V.; Bakwin, P.S.; Yi, C.X.; et al. Carbon exchange and venting anomalies in an upland deciduous forest in northern Wisconsin, USA. *Agric. For. Meteorol.* **2004**, *126*, 271–295. [CrossRef]

41. Desai, A.R.; Bolstad, P.V.; Cook, B.D.; Davis, K.J.; Carey, E.V. Comparing net ecosystem exchange of carbon dioxide between an old-growth and mature forest in the upper Midwest, USA. *Agric. For. Meteorol.* **2005**, *128*, 33–55. [CrossRef]

42. Rey, A.; Pegoraro, E.; Tedeschi, V.; De Parri, L.; Jarvis, P.G.; Valentini, R. Annual variation in soil respiration and its components in a coppice oak forest in central Italy. *Glob. Chang. Biol.* **2002**, *8*, 851–866. [CrossRef]

43. Curiel, Y.J.; Janssens, I.A.; Carrara, A.; Meiresonne, L.; Ceulemans, R. Interactive effects of temperature and precipitation on soil respiration in a temperate maritime pine forest. *Tree Physiol.* **2003**, *23*, 1263–1270.

44. Flanagan, L.B.; Johnson, B.G. Interacting effects of temperature, soil moisture and plant biomass production on ecosystem respiration in a northern temperate grassland. *Agric. For. Meteorol.* **2005**, *130*, 237–253. [CrossRef]

45. Raich, J.W.; Schlesinger, W.H. The global carbon dioxide flux in soil respiration and its relationship to vegetation and climate. *Tellus B* **1992**, *44*, 81–99. [CrossRef]

46. Davidson, E.A.; Belt, E.; Boone, R.D. Soil water content and temperature as independent or confounded factors controlling soil respiration in a temperate mixed hardwood forest. *Glob. Chang. Biol.* **1998**, *4*, 217–227. [CrossRef]

47. Curiel, Y.J.; Janssens, I.A.; Carrara, A.; Ceulemans, R. Annual Q_{10} of soil respiration reflects plant phonological patterns as well as temperature sensitivity. *Glob. Chang. Biol.* **2004**, *10*, 161–169. [CrossRef]

48. Cao, G.M.; Tang, Y.H.; Mo, W.H.; Wang, Y.S.; Li, Y.N.; Zhao, X.Q. Grazing intensity alters soil respiration in an alpine meadow on the Tibetan plateau. *Soil Biol. Biochem.* **2004**, *36*, 237–243. [CrossRef]

49. Wu, J.B.; Guan, D.X.; Wang, M.; Pei, T.F.; Han, S.J.; Jin, C.J. Year-round soil and ecosystem respiration in a temperate broad-leaved Korean Pine forest. *For. Ecol. Manag.* **2006**, *223*, 35–44. [CrossRef]

50. Li, H.J.; Yan, J.X.; Yue, X.F.; Wang, M.B. Significance of soil temperature and moisture for soil respiration in a Chinese mountain area. *Agric. For. Meteorol.* **2008**, *148*, 490–503. [CrossRef]

51. Granier, A.; Ceschia, E.; Damesin, C.; Dufrêne, E.; Epron, D.; Gross, P.; Lebaube, S.; Le Dantec, V.; Le Goff, N.; Lemoine, D.; et al. The carbon balance of a young Beech forest. *Funct. Ecol.* **2000**, *14*, 312–325. [CrossRef]

52. Simmons, J.A.; Fernandez, I.J.; Briggs, R.D.; Delaney, M.T. Forest floor carbon pools and fluxes along a regional climate gradient in Maine, USA. *For. Ecol. Manag.* **1996**, *84*, 81–95. [CrossRef]

53. Yuji, K.; Mayuko, J.; Masako, D.; Yoshiaki, G.; Koji, T.; Takafumi, M.; Yoichi, K.; Shinji, K.; Motonori, O.; Noriko, M.; et al. Biometric and eddy-covariance-based estimates of carbon balance for a warm-temperate mixed forest in Japan. *Agric. For. Meteorol.* **2008**, *148*, 723–737.

54. Wang, M.; Liu, Y.Q.; Hao, Z.Q.; Wang, Y.S. Respiration rate of broad-leaved Korean pine forest ecosystem in Changbai Mountain. *Chin. J. Appl. Ecol.* **2006**, *17*, 1789–1795. (In Chinese, with English Abstract)

55. Tan, Z.H.; Zhang, Y.P.; Yu, G.R.; Sha, L.Q.; Tang, J.W.; Deng, X.B.; Song, Q.H. Carbon balance of a primary tropical seasonal rain forest. *J. Geophys. Res.* **2010**, *115*, 411–454. [CrossRef]

56. Law, B.E.; Thornton, P.E.; Irvine, J.; Anthoni, P.M.; Van Tuyl, S. Carbon storage and fluxes in ponderosa pine forests at different developmental stages. *Glob. Chang. Biol.* **2001**, *7*, 755–777. [CrossRef]

57. Amthor, J.S. Higher plant respiration and its relationships to photosynthesis. *Ecophysiol. Photosynth.* **1993**, *100*, 71–102.

58. Wang, M.; Ji, L.Z.; Li, Q.R.; Xiao, D.M.; Liu, H.L. Stem respiration of Pinus koraiensis in Changbai Mountain. *Chin. J. Appl. Ecol.* **2005**, *16*, 7–13. (In Chinese, with English Abstract)

59. Ryan, M.G.; Gower, S.T.; Hubbard, R.M.; Waring, R.H.; Gholz, H.L.; Cropper, W.P., Jr.; Running, S.W. Woody tissue maintenance respiration of four conifers in contrasting climates. *Oecologia* **1995**, *101*, 133–140. [CrossRef] [PubMed]

60. Ma, Y.E.; Xiang, W.H.; Lei, P.F. Stem respiration and its controlling factors in forest ecosystems. *Chin. J. Plant Ecol.* **2007**, *31*, 403–412. (In Chinese, with English Abstract)

61. Gough, C.M.; Vogel, C.S.; Kazanski, C.; Nagel, L.; Flower, C.E.; Curtis, P.S. Coarse woody debris and the carbon balance of a north temperate forest. *For. Ecol. Manag.* **2007**, *244*, 60–67. [CrossRef]

62. Bond-Lamberty, B.; Wang, C.; Gower, S.T. Annual carbon flux from woody debris for a boreal black spruce fire chronosequence. *J. Geophys. Res.* **2002**, *108*, 8220–8230. [CrossRef]

63. Manies, K.L.; Harden, J.W.; Bond-Lamberty, B.; O'Neill, K.P. Woody debris along an upland chronosequence in boreal Manitoba and its impact on long-term carbon storage. *Can. J. For. Res.* **2005**, *35*, 472–482. [CrossRef]

64. Progar, R.A.; Schowalter, T.D.; Freitag, C.M.; Morrell, J.J. Respiration from coarse woody debris as affected by moisture and saprotroph functional diversity in Western Oregon. *Oecologia* **2000**, *124*, 426–431. [CrossRef] [PubMed]

65. Wang, C.K.; Bond-Lamberty, B.; Gower, S.T. Environmental controls on carbon dioxide flux from black spruce coarse woody debris. *Oecologia* **2002**, *132*, 374–381. [CrossRef] [PubMed]

66. Wu, J.B.; Guan, D.X.; Han, S.J.; Pei, T.F.; Shi, T.T.; Zhang, M. Respiration of fallen trees of Pinus koraiensis and Tilia amurensis in Changbai Mountain, northeastern China. *J. Beijing For. Univ.* **2008**, *30*, 14–19. (In Chinese, with English Abstract)

67. Zhang, S.B.; Zheng, Z.A. Preliminary research on respiration of woody debris of hollow-bearing tree in the montane moist evergreen broad-leaved forest of Ailao Mountain, Yunnan, China. *J. Mt. Sci. Engl.* **2008**, *26*, 300–307.

68. Weedon, J.T.; Cornwell, W.; Cornelissen, J.H.C.; Zanne, A.E.; Wirth, C.; Coomes, D.A. Global meta-analysis of wood decomposition rates, a role for trait variation among tree species? *Ecol. Lett.* **2009**, *12*, 45–56. [CrossRef] [PubMed]

69. Shorohova, E.; Ekaterina, K. Influence of the substrate and ecosystem attributes on the decomposition rates of coarse woody debris in European boreal forests. *For. Ecol. Manag.* **2014**, *315*, 173–184. [CrossRef]

70. Rice, A.H.; Pyle, E.H.; Saleska, S.; Hutyra, L.; Palace, M.; Keller, M.; De Camargo, P.B.; Portilho, K.; Marques, D.F.; Wofsy, S.C. Carbon balance and vegetation dynamics in an old growth Amazonian forest. *Ecol. Appl.* **2004**, *14*, 55–71. [CrossRef]

71. Cook, B.D.; Bolstad, P.V.; Martin, J.G.; Heinsch, F.A.; Davis, K.J.; Wang, W.G.; Desai, A.R.; Teclaw, R.M. Using light-use and production efficiency models to predict photosynthesis and net carbon exchange during forest canopy disturbance. *Ecosystems* **2008**, *11*, 26–44. [CrossRef]

72. Nagy, Z.; Pintér, K.; Pavelka, M.; Darenová, E.; Balogh, J. Carbon fluxes of surfaces vs. ecosystems: Advantages of measuring eddy covariance and soil respiration simultaneously in dry grassland ecosystems. *Biogeosciences* **2011**, *8*, 2523–2534. [CrossRef]

73. Goulden, M.L.; Munger, J.W.; Fan, S.M.; Daube, B.C.; Wofsy, S.C. Measurements of carbon sequestration by long-term Eddy covariance: Methods and a critical evaluation of accuracy. *Glob. Chang. Biol.* **1996**, *2*, 169–182. [CrossRef]

74. Goulden, M.L.; Wofsy, S.C.; Harden, J.W.; Trumbore, S.E.; Crill, P.; Gower, S.T.; Fries, T.; Daube, B.C.; Fan, S.M.; Sutton, D.J.; et al. Sensitivity of boreal forest carbon balance to soil thaw. *Science* **1998**, *279*, 214–217. [CrossRef] [PubMed]

75. Grace, J.; Malhi, Y.; Lloyd, J.; Mcintyre, J.; Miranda, A.C.; Meir, P.; Miranda, H.S. The use of eddy covariance to infer the net carbon dioxide uptake of Brazilian rain forest. *Glob. Chang. Biol.* **1996**, *2*, 209–217. [CrossRef]

76. Riederer, M.; Serafimovich, A.; Foken, T. Net ecosystem CO_2 exchange measurements by the closed chamber method and the eddy covariance technique and their dependence on atmospheric conditions. *Atmos. Meas. Tech.* **2014**, *7*, 1057–1064. [CrossRef]

77. Curtis, P.S.; Hanson, P.J.; Bolstad, P.; Barford, C.; Randolph, J.C.; Schmid, H.P.; Wilson, K.B. Biometric and eddy-covariance based estimates of annual carbon storage in five eastern north american deciduous forests. *Agric. For. Meteorol.* **2002**, *113*, 3–19. [CrossRef]

78. Campioli, M.; Malhi, Y.; Vicca, S.; Luyssaert, S.; Papale, D.; Peñuelas, J.; Reichstein, M.; Migliavacca, M.; Arain, M.A.; Janssens, I.A. Evaluating the convergence between eddy-covariance and biometric methods for assessing carbon budgets of forests. *Nat. Commun.* **2016**, *7*, 13717. [CrossRef] [PubMed]

79. Ehman, J.L.; Schmid, H.P.; Grimmond, C.S.B.; Randolph, J.C.; Hanson, P.J.; Wayson, C.A.; Cropley, F.D. An initial intercomparison of micrometeorological and ecological inventory estimates of carbon exchange in a mid-latitude deciduous forest. *Glob. Chang. Biol.* **2002**, *8*, 575–589. [CrossRef]

80. Clark, D.A.; Brown, S.; Kicklighter, D.W.; Chambers, J.Q.; Thomlinson, J.R.; Ni, J. Measuring net primary production in forests: Concepts and field methods. *Ecol. Appl.* **2001**, *11*, 356–370. [CrossRef]

81. Houghton, R.A.; Skole, D.L.; Nobre, C.A.; Hackler, J.L.; Lawrence, K.T.; Chomentowski, W.H. Annual fluxes of carbon from deforestation and regrowth in the Brazilian Amazon. *Nature* **2000**, *403*, 301–304. [CrossRef] [PubMed]

82. Fang, J.Y.; Chen, A.P.; Peng, C.H.; Zhao, S.Q.; Ci, L.J. Changes in forest biomass carbon storage in China between 1949 and 1998. *Science* **2001**, *292*, 2320–2322. [CrossRef] [PubMed]

83. Gough, C.M.; Vogel, C.S.; Schmid, H.P.; Su, H.B.; Curtis, P.S. Multi-year convergence of biometric and meteorological estimates of forest carbon storage. *Agric. For. Meteorol.* **2008**, *148*, 158–170. [CrossRef]

84. Vicca, S.; Luyssaert, S.; Peñuelas, J.; Campioli, M.; Chapin, F.S.; Ciais, P.; Heinemeyer, A.; Högberg, P.; Kutsch, W.L.; Law, B.E.; et al. Fertile forests produce biomass more efficiently. *Ecol. Lett.* **2012**, *15*, 520–526. [CrossRef] [PubMed]

 *forests*

Article

Carbon Mass Change and Its Drivers in a Boreal Coniferous Forest in the Qilian Mountains, China from 1964 to 2013

Shu Fang [1,2], Zhibin He [1,*], Jun Du [1], Longfei Chen [1], Pengfei Lin [1,2] and Minmin Zhao [1,2]

[1] Linze Inland River Basin Research Station, Chinese Ecosystem Research Network, Key Laboratory of Eco-hydrology of Inland River Basin, Northwest Institute of Eco-Environment and Resources, Chinese Academy of Sciences, Lanzhou 730000, China; fangs@lzb.ac.cn (S.F.); dujun159@126.com (J.D.); chenlongfei_mail@163.com (L.C.); linpengfei@lab.ac.cn (P.L.); zhaomm@lab.ac.cn (M.Z.)

[2] University of Chinese Academy of Sciences, Beijing 100049, China

* Correspondence: hzbmail@lzb.ac.cn; Tel.: +86-136-6930-4220

Received: 16 November 2017; Accepted: 22 January 2018; Published: 25 January 2018

Abstract: Carbon storage of mountain forests is vulnerable to climate change but the changes in carbon flux through time are poorly understood. Moreover, the relative contributions to carbon flux of drivers such as climate and atmospheric CO_2 still have significant uncertainties. We used the dynamic model LPJ-GUESS with climate data from twelve meteorological stations in the Qilian Mountains, China to simulate changes in carbon mass of a montane boreal forest, and the influence of temperature, precipitation, and CO_2 concentration from 1964 to 2013 on carbon flux. The results showed that the carbon mass has increased 1.202 kg/m^2 from 1964 to 2013, and net primary productivity (NPP) ranged from 0.997 to 1.122 kg/m^2/year. We concluded that the highest carbon mass proportion for this montane boreal forest was at altitudes 2700–3100 m (proportion of ecosystem carbon was between 93–97%), with maximum carbon density observed at 2700–2900 m. In the last 50 years, the increase in precipitation and in CO_2 concentration is expected to increase carbon mass and NPP of *Picea crassifolia* Kom. (Pinaceae) (Qinghai spruce). The effect of temperature on NPP was positive but that on carbon mass was not clear. The increase in CO_2 concentration over the past 50 years was a major contributor to the increase in carbon storage, and drought was the foremost limiting factor in carbon storage capacity of this montane boreal forest. *Picea crassifolia* forest was vulnerable to climate change. Further studies need to focus on the impact of extreme weather, especially drought, on carbon storage in *Picea crassifolia* forests.

Keywords: carbon mass; NPP; *Picea crassifolia*; climate change

1. Introduction

Climate change influences the carbon and water cycles in mountain areas [1]. Forests in mountain areas provide important ecological and socio-economic services; however, carbon storage is vulnerable to the effects of climate change and may decrease its ecological service function over time [2,3]. Research on the response of mountain ecosystems, especially of alpine forests and tree species, to climate change, is lacking [1,2]. Additionally, the impacts of climate change on montane forests are examined distinctively with dendroclimatic techniques [2–4]. High mountain boreal forests, which have numerous organic pools stored aboveground and in the permafrost, play an important role in regional carbon budgets and are exposed to rapid climate change; however, research on carbon pools at high elevations and controls on carbon flux through time lag behind [5,6] due primarily to insufficient data [7]. Ecosystem models are useful tools for describing and quantifying water, matter, and momentum fluxes between the biosphere and the atmosphere [8]. Using biomass models to obtain mass data can facilitate simulation of forest carbon dynamics in areas where such data are lacking, while analyses of the montane boreal forest biome will improve the accuracy of global carbon models [3].

Climate change is primarily related to changes since the 1950s in solar radiation, temperature, and precipitation that have not been observed before [9]. Climate change and related disturbances may substantially impact the species ranges, population sizes, and extinction risks in mountain forests, and even shift the carbon sink activity of a forest to a net carbon source [10–12]. In a simulation of a species distribution in Mediterranean mountains, the cold-adapted species decreased in significance, and the species range and communities changed under global warming [13]. Extreme climates may change plant intrinsic characteristics and increase the possibility of mountain ecosystem shifts from forest to shrubland or grassland [14]. These climate change-induced species conversions will result in a long-term loss of carbon stock in mountain forest ecosystems [15]. One stark example of this was observed with mountain pine beetle in western North America; this insect outbreak, caused by climate change, compromised the ability of forests to take up and store carbon [12].

The response of mountain forests to climate change depends heavily on relative changes in temperature, precipitation, CO_2, and their interaction effect [10]; the relative contributions of individual historical drivers can be assessed by high-resolution climate data and modeling [16,17]. Sensitivity analyses of climate factors focused on tropical and temperate areas. However, analyses of the driving factors of carbon storage changes in recent years at high latitudes and in high elevation areas are still lacking [18]. The sensitivity to climate change of carbon storage in mountain forests is based on initial climatic variability; thus, increased mean annual temperature at high elevations results in sensitivity of tree growth and changes in forest productivity [3]. Examined by three ecological models (LPJ-GUESS, ForClim, LandClim), carbon storage across all elevations revealed a lower sensitivity than other ecosystem services to a 2 °C warming [19]. However, under the influence of global warming, mountain forests at high altitudes exhibited positive growth, whereas drought stress led to a negative effect on carbon mass at lower altitudes [4]. The impacts of climate change on mountain forests were also closely related to the frequency and intensity of extreme weather, especially precipitation-related events [5]. Extreme warming events may influence mountain boreal plant activity, plant litter production, soil moisture, and insect life cycles, while extreme weather events such as precipitation extremes and severe storms may lead to damage and may influence the recovery capacity of mountain boreal forest ecosystems [6]. Uncertainties exist as to the main climatic controls on carbon changes in long-term simulations of carbon mass in mountain forests [9,10]. Investigating carbon flux sensitivity to climate change will reduce the uncertainty of the past factors of climate change and help guide forest management. For a better understanding of the effects of climate change on boreal forests, it is necessary to quantify the driving parameters of carbon changes in boreal forests during the last 50 years.

The Qilian Mountains, located in the northwestern part of China at the north-eastern edge of the Tibetan Plateau, are typical arid mountains, and the source of key inland rivers; these mountains are sensitive to climate change [20]. The average temperature rise is 80% higher than the national rate (0.14 °C/10a), and extreme climate and precipitation events occurred more frequently here in the past 50 years than in other parts of China [21]. Under climate change, environmental degradation, soil erosion, and loss of biodiversity have increased, and vegetation cover and growth in the Qilian Mountains have changed significantly in recent years [22]. Montane Boreal forest is the dominant forest type, and it plays important roles in the hydrology and biogeochemistry of the area due to high altitudes (2000–5500 m) and minimal human-induced ecological deterioration [23]. Between 1957 and 2007, temperature increased by 0.29 °C/10a and precipitation by 5.5 mm/10a, resulting in a shift of the tree line of the montane boreal forest in this area to higher elevations [24]. Simple regression models of phenological dynamics in this montane boreal forest revealed that the length of the growing season can be expected to increase during the next two decades [25]. This prompted new research efforts to determine the relationships between montane boreal forest carbon dynamics and climate change. Ecosystem stability of the montane boreal forest in the Qilian Mountains supports downstream economy and production because the forest serves as the most important water conservation entity in the area. Better understanding of the role of climate and atmospheric CO_2 in determining carbon flux in this montane boreal forest, and the relationship between the NPP and biomass is critical for guiding ecology policy to optimize carbon storage.

To address this knowledge need, we used climate and CO_2 concentration data to reconstruct carbon storage in living biomass for the montane boreal forest in the Qilian Mountains, China from 1964–2013. With this, we examined the carbon stock in this forest type at different elevations and its carbon flows during the last 50 years. Subsequently, we used the LPJ-GUESS model to explore the sensitivity of forest carbon stocks in this region to temperature, precipitation, CO_2, and extreme conditions to determine the main past factor controlling the change in carbon flux in the last 50 years.

2. Materials and Models

2.1. Area Description and Data Collection

The Qilian Mountains are one of the major arid mountain ranges in northwestern China, located in the northeastern Tibetan Plateau, between 93°33′36″–103°54′00″ E and 35°50′24″–40°01′12″ N. The distribution of vegetation is strongly controlled by altitude and can be separated into five types (from low to high altitude): desert steppe, forest steppe, sub-alpine shrubby meadow, alpine cold desert, and ice/snow zone. In the forest steppe, the montane boreal forest is located at high altitudes where the level of human impact is low [26]. The dominant boreal species, *Picea crassifolia*, grows on shady and partly-shady north slopes at altitudes ranging from 2600 to 3400 m, and accounts for as much as 76% of the total forested area in the forest steppe zone [27]. In total, twelve sampling sites were selected across an elevation gradient from 2300–3500 m at intervals of 200 m to analyze the carbon change of montane boreal forest by altitude.

We selected twelve meteorological stations located within the tree-growing areas to extract the climate parameters (Figure 1). In an earlier study of the spatial distribution of *Picea crassifolia* biomass and carbon storage in the Qilian Mountains, the average carbon content was calculated as 0.52 [28]; this was similar to the results based on a synthesis of carbon contents for temperate and boreal conifer wood (n = 36) of 50.8 ± 0.6% [29]. In this study, we used a carbon content of 0.52 for *Picea crassifolia*.

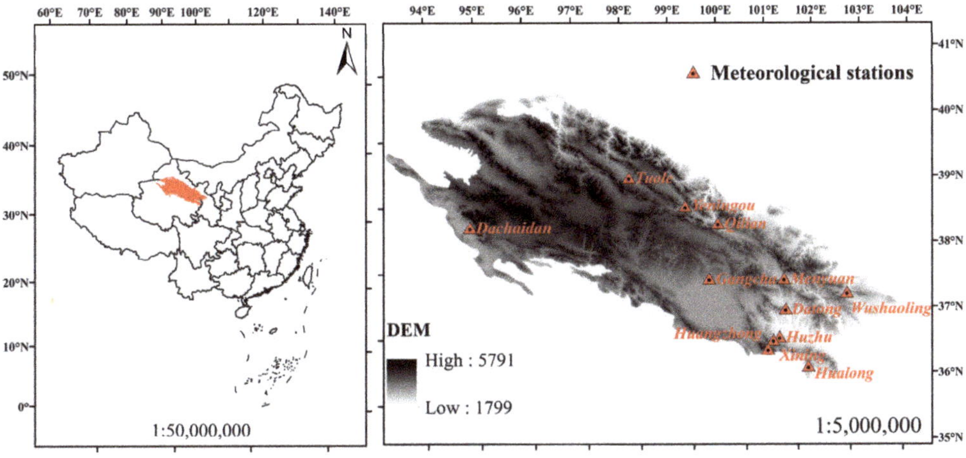

Figure 1. Study area and locations of meteorological stations used in the study in the Qilian Mountains.

The climate data, including mean monthly temperature, monthly precipitation, mean monthly daily proportion of sunshine hours, and monthly rainy days, were obtained from the climate data repository of Chinese meteorological stations (http://data.cma.cn/site/index.html), and of Qinghai Province (Table 1). Atmospheric CO_2 concentrations for 1964–2013 were derived from the atmospheric carbon dioxide mixing ratios from NOAA (National Oceanic and Atmospheric Administration) ESRL (Earth System Research Laboratory) Carbon Cycle Cooperative Global Air Sampling Network (http://www.esrl.noaa.gov/gmd/ccgg/mbl/data.php).

Table 1. Site information.

Groups	Name	Level	Longitude (°)	Latitude (°)	Altitude (m)
2300–2500 m	Xining	Provincial-Qinghai	101.45	36.44	2295
	Huzhu	Provincial-Qinghai	101.57	36.49	2480
2500–2700 m	Datong	Provincial-Qinghai	101.67	36.92	2587
	Huangzhong	Provincial-Qinghai	101.35	36.3	2668
2700–2900 m	Qilian	National-China	100.25	38.18	2787
	Hualong	Provincial-Qinghai	102.15	36.06	2835
	Menyuan	National-China	101.62	37.38	2850
2900–3100 m	Wushaoling	National-China	102.52	37.12	3045
3100–3300 m	Dachaidan	National-China	95.37	37.85	3173
	Yeniugou	National-China	99.58	38.42	3180
3300–3500 m	Gangcha	National-China	100.13	37.33	3302
	Tuole	National-China	98.42	38.8	3367

2.2. The LPJ-GUESS Model

LPJ-GUESS is a validated vegetation carbon dynamics model [30]. The basic use of the model is to simulate carbon balance of different tree species and to predict changes in carbon pools and fluxes under global climate change [31,32]. A LPJ-GUESS simulation of the spatial and temporal patterns of carbon fluxes associated with regrowth after agricultural abandonment indicated that semi-arid regions, where carbon balance was strongly associated with both precipitation and temperature, were key to the understanding and predicting of the global carbon cycle [33]. Further, changes in the spatial patterns of wildfires, estimated by LPJ-GUESS, were critical to the proposal for a more reasonable climate policy [34]. The model can simulate and predict the responses of plant water fluxes to elevated CO_2 at leaf and stand scales [35]. In addition to natural vegetation, agricultural crop production and crop responses to climate change were analyzed in an effort to increase food security [36]. In China, the model has been applied in subtropical, temperate, and mountain zones [37–39], and to the entire country [40]. Carbon simulation by LPJ-GUESS using an individual climate factor demonstrated carbon release by management, and CO_2 as the most important driver for carbon change in Lady Park Wood in the last 20 years [16].

Plant functional types (PFTs), which can simplify species diversity of vegetation, represent groups of species with similar functional traits. Among forest carbon dynamics simulation models, both, the individual-based model GUESS (General Ecosystem Simulator) [30], and the area-based dynamic global vegetation model LPJ (Lund-Potsdam-Jena) [41], can be successful in predicting the carbon flux of PFTs. Subsequently, a new vegetation dynamics model, LPJ-GUESS, was developed by combining the two other models [30].

Two model frameworks exist, the cohort and population. We employed the "cohort mode" in this study. In the "cohort mode", individual trees are distinguished, but are identical within each cohort (age class) [30]. Population processes and disturbances are modeled stochastically, and stand characteristics are averaged over 100 patches of 0.1 ha, representing "random samples" of the simulated stand. The model is driven by short-wave radiation (photosynthetically active light), temperature, precipitation, CO_2 concentration of the air, and soil conditions. The CO_2 influences assimilation rate, and soil conditions modify plant water uptake; soil data are already provided in the model. Species information includes the physiological characteristics of PFTs, such as prescribed allometric relationships. Based on physiology, morphology, phenology, and on the response to disturbance and to bio-climatic limiting factors, the model defines ten PFTs, of which 8 are woody, and two are herbaceous [42]. We used the full set of PFTs and added a new PFT for *Picea crassifolia* to reproduce the current stage of the vegetation.

2.3. Model Forcing and Simulation Protocol

The simulation normally follows two or three phases and begins with "bare ground", which means that the modelled area is bare, with no vegetation present. The first phase of the simulation is known as the "spin-up"; in this phase, input data are normally based on the first few years of available (historical) data. During the subsequent, "historical" phase, the model uses "observed" climate and CO_2 data as input. We chose two phases, "spin-up" and "historical" to run the model for the last fifty years. Climate data for an initial 300-year "spin-up" phase were not available; such "spin-up" equilibrates the initial vegetation and carbon pools with climate at the beginning of the study period. The model was first "spun-up" for 300 years, recycling the observed time series from 1964–1993 [30,43]. The "historical" period then ran from 1964 to 2013; the simulation time interval for the LPJ-GUESS model is five years.

In a study to investigate the ability of LPJ-GUESS to reproduce features of real vegetation, the model was demonstrated to not require site-specific calibration and could be used to simulate vegetation dynamics on a regional basis or under past or future climates and atmospheric CO_2 levels for reparameterization becauase plant growth is modeled mechanistically [31]. Many of the parameters of species are decided by its PFTs, such as whether it is needle-leaved or broad-leaved; or whether it is boreal or temperate species. The species-specific parameters such as tree longevity are defined in the article. We referred to the relevant parameters for the boreal forest to define the parameters of *Picea crassifolia* [31,44]. We specified the mean length of the life of foliage at 11.8 years, as determined in a field investigation [45], and tree longevity at 250 years, as described in the literature [46]. Data used for biomass calculations for the Qilian Mountains were obtained from the literature [28,47–49] and were used to verify the accuracy of those that were calculated.

2.4. Sensitivity Analysis of Temperature, CO_2, and Extreme Whether

Sensitivity analysis is an important aspect of evaluating the resilience of ecosystems to climate change and can be combined with models to measure the effect of changing input parameters under some climate scenarios and extreme weather conditions [50,51]. Based on the simulation of LPJ-GUESS, further research was conducted to determine the relative effect of climate variables and CO_2 concentration on the carbon flux of montane boreal forests in Qilian Mountains for the period 1964–2013.

The temperature increased during the simulation period, but this trend could not be fit by a linear model. Further, the trend in precipitation was not clear. Thus, we first examined the climate effect by simulating the temperature, precipitation, and CO_2 independently of each other. Using these climate simulations as a baseline, the uncertainties of these parameters were then compared. The influence of the temperature was observed for an increase or decrease of 1 and 2 °C; precipitation was changed multiplicatively, because it is a zero-based variable, by increasing or decreasing by 10% and 20%. The influence of CO_2 was examined by removing its trend and increasing the value by 50 ppm and 100 ppm.

Because the climate variables are strongly related to each other, a comparison of the effects of climate change on carbon flux based on changing a single parameter in the data was incomplete [6,16]. Hence, to preserve the relationship of temperature, precipitation, and radiation, we represented extreme weather as follows to determine the impact of climate factors on carbon mass. Local weather in the study area was ranked in terms of temperature or precipitation levels (annual mean temperature from April to October, and annual mean precipitation from May to September), and the top five warmest, coldest, wettest, and driest years; these were then cycled through the model repeatedly to simulate an extreme climate.

3. Results

3.1. Changes in Carbon Mass with Altitude

The monthly mean temperature and precipitation in the simulated period followed normal distribution. Warm periods were concentrated in April to October, and high precipitation was concentrated in May to September; almost all of the stations exhibited the same trends. Mean annual temperature at the twelve stations increased by 0.29–3.69 °C between 1964 and 2013, with an average increase of 0.73 °C/10a. Differences in annual precipitation among altitudes were not apparent.

Carbon stocks of the montane boreal forest were concentrated at 2700–3100 m and exhibited a single peak of 11.787 kg/m^2 at 2700–2900 m in 2013, with the average carbon mass at 2700–3100 m of 10.503 kg/m^2. The calculation of the biomass of *Picea crassifolia* was different for the different measurement methods and research areas, but the results of our simulation were within the range of other studies (Table 2). The proportion of montane boreal forest was >28% at all altitudes. Specifically, montane boreal forest accounted for 30% at 2300 to 2700 m, increasing to 93% at 2700–2900 m and 97% at 2900–3100 m, declining to 37%, and again reaching 49% at 3300–3500 m (Figure 2). The carbon stored in the ecosystem at 2300–2700 m was 7.068–7.591 kg·C/m^2, declining to about 0.242–0.645 kg·C/m^2 at 3100–3500 m.

Table 2. Carbon mass calculations for *Picea crassifolia* in the literature.

Method	Research Area	Carbon Mass	Sources
LPJ-GUESS	Qilian Mountains	10.924 (kg/m^2)	This study
Sample measured	Sunan County	28.250 (kg/m^2)	[48]
Sample measured	Qilian Mountains	20.920 (kg/m^2)	[28]
Vegetation survey and Tree-ring research	Pailugou Watershed	12.861 (kg/m^2)	[49]
Sample measured	Haxi forest farm	13.290 (kg/m^2)	[49]
Model simulation	Qilian Mountains	16.980 (kg/m^2)	
Sample measured	north-eastern edge Qilian Mountains	8.270(kg/m^2)	[47]

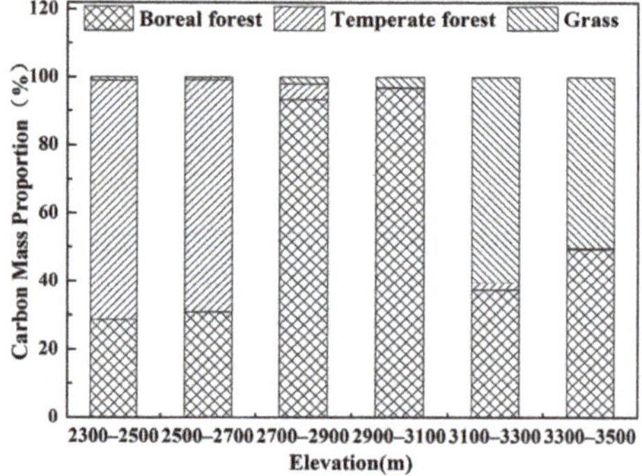

Figure 2. Carbon mass proportion at different elevations for twelve meteorological stations. Data are the means of simulated values for each range in elevation.

We investigated the NEE (Net ecosystem exchange) and carbon mass change in the mountain boreal forest ecosystem (at 2700–3100 m where the boreal forest accounted for the largest carbon mass) (Figure 3). In the model, a positive value of NEE indicated that the ecosystem was a carbon source, while a negative value indicated a carbon sink. Boreal forest was a carbon sink before the 1980s; sink strength at elevation 2700–2900 m declined, until the sink activity became a carbon source after 1980, and it became a relatively strong source in the 1990s. However, the strength of the carbon source weakened and the forest ecosystem at 2900–3100 m became a carbon sink again after the year 2000. We analysed the change in carbon mass during 1964–2013 and found the there was an upward trend for this montane boreal forest. The biomass of *Picea crassifolia* forest below 3100 m decreased in the study period, with the greatest decline of 2.595 kg/m^2 at 2700–3100 m, and least of 0.125 kg/m^2 at 2300–2500 m. The carbon mass increased by 0.035 kg/m^2 at 3100–3300 m, and by 0.256 kg/m^2 at 3300–3500 m. As the NEE indicated, the decreasing trend of carbon mass for montane boreal forest at 2700–2900 m stabilized in the 2000s and even exhibited a slight increase; carbon mass at 2900–3100 m was stable before the 1990s, and increased after that.

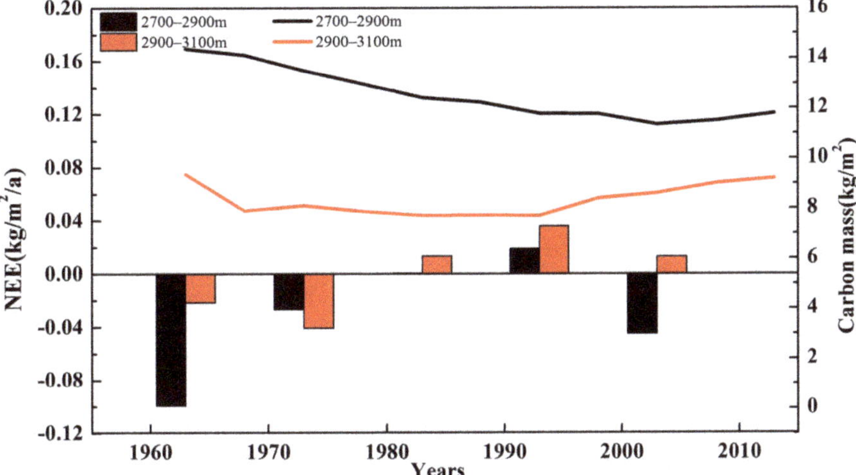

Figure 3. The carbon flows change in montane boreal forest ecosystem from 1964–2013. kg/m^2/a: kg/m^2/year.

3.2. The Sensitivity Analysis of Carbon Mass of Montane Boreal Forest to Climate Variables and CO$_2$

The simulation showed that the carbon carrying capacity of *Picea crassifolia* increased in the study period, given no changes to any variable. Mean carbon mass increased by 1.202 kg/m^2 from 1964–2013, and the 5-year (simulation time interval for the LPJ-GUESS model) mean NPP increased by 0.023 kg/m^2/year to between 0.997 and 1.122 kg/m^2/year in the simulation period. Subsequently, each of the climatic variables was changed one at a time.

Following a change in temperature, the trend in NPP did not change, but the NPP value was higher than before (Figure 4). The effect of lower temperature on NPP was greater than that of higher temperature. When the temperature increased by 1 °C, 5-year mean NPP increased by 0.090 kg/m^2/year; when the temperature increased by 2 °C, mean NPP increased by 0.148 kg/m^2/year per five years. When the temperature decreased by 1 °C, mean NPP decreased 0.113 kg/m^2/year per five years; when the temperature decreased by 2 °C, mean NPP decreased by 0.292 kg/m^2/year per five years. However, the change in carbon stocks was not consistent with NPP when the temperature changed; increased or decreased temperature appeared to suppress carbon stocks of the montane boreal forest. Mean carbon stock decreased between 1964–2013 by 0.768 kg/m^2 per five years; the increments decreased from 12.37% to 6.32% as the temperature increased by 1 °C. As the temperature increased by 2 °C, mean carbon stock decreased by 0.893 kg/m² per five years, and carbon mass was reduced by 8.24%. With the temperature decrease of 1 °C, carbon stock increased by 0.026 kg/m² per five years and the increments increased to 13.28%. With the temperature decrease of 2 °C, mean carbon stock decreased by 1.307 kg/m² per five years, and carbon mass decreased by 10.11% during the last 50 years.

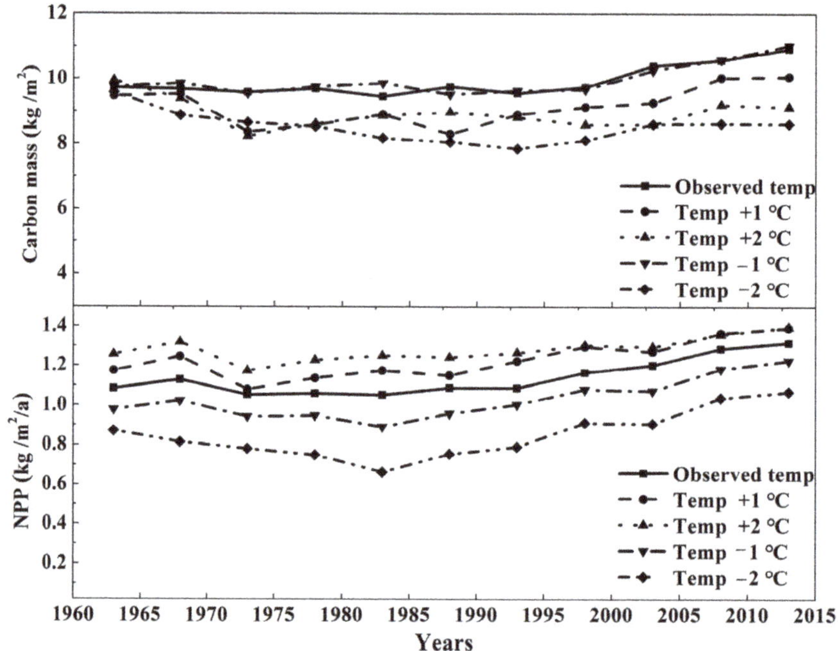

Figure 4. LPJ-GUESS output showing the effects on stored carbon (top panel) and NPP (bottom panel) of changing the temperature only. Temperature was modified by adding or subtracting 1 (or 2) °C to each daily climate value. Lines represent the means of the simulations. Temp: the abbreviation of temperature.

The trend in NPP changed over the course of simulation following changes in precipitation (Figure 5); the more precipitation, the higher the NPP, and the less precipitation, the lower the NPP. The impact of reduced precipitation was greater than that of increased precipitation. A 20% increase in precipitation resulted in a mean 5-year NPP increase of 0.024 kg/m²/year, while a 20% decrease reduced NPP by 0.080 kg/m²/year per five years.

The changes in carbon stock were consistent with NPP. Mean carbon stocks increased 0.242 kg/m² per five years as precipitation increased by 10%, and 1.165 kg/m² per five years as the precipitation increased by 20%. As precipitation decreased by 10 and 20%, mean carbon stocks decreased 1.517 and 2.969 kg/m² per five years, respectively. The increase in carbon mass resulting from increased precipitation was 24.39%, and 24.93%, respectively, (basic value 12.37%), and carbon mass decreased 19.17%, and 34.01%, respectively, if precipitation decreased.

Mean NPP decreased when the CO_2 concentration remained unchanged (de-trended) and increased when the CO_2 concentration increased (Figure 6). Carbon mass decreased when the CO_2 concentration remained unchanged, but it increased as the CO_2 concentration levels rose. Overall, the CO_2 concentration had an increasingly positive effect as atmospheric levels rose. Carbon mass decreased by 0.240 kg/m² for the simulation years if CO_2 concentrations did not increase after 1964. Mean carbon storage increased by 0.746 and 0.866 kg/m^2 per five years for CO_2 concentration increases of 50 and 100 ppm. De-trended CO_2 supressed carbon mass by about 2.45%, and increased CO_2 promoted carbon storage from 12.37% to 18.98%, and to 24.74% in response to 50 and 100 ppm, respectively.

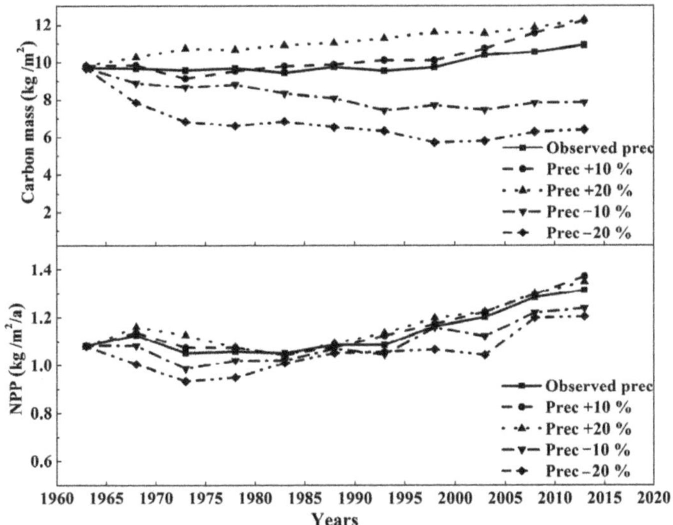

Figure 5. LPJ-GUESS output showing the effects on stored carbon (top panel) and NPP (bottom panel) of changing precipitation only. Precipitation was changed multiplicatively because it is a zero-based variable. Lines represent the means of the simulations. Prec: the abbreviation of precipitation.

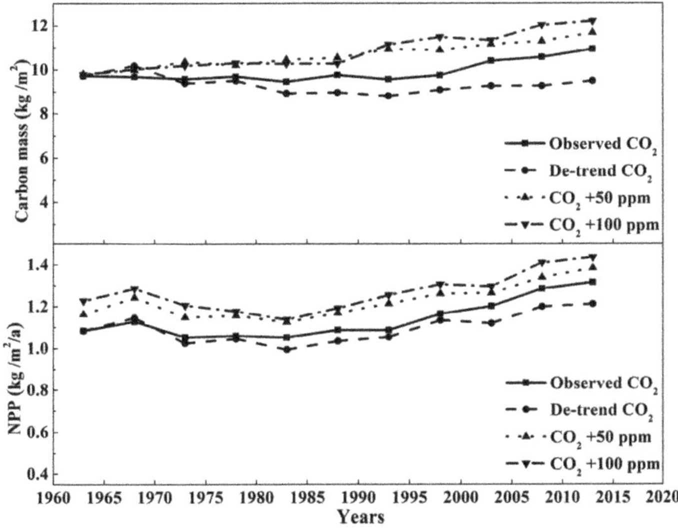

Figure 6. LPJ-GUESS output showing the effect on stored carbon (top panel) and NPP (bottom panel) of changing CO_2 concertration only. CO_2 concertration was de-trended by using 1964 levels throughout and was modified by adding values to each year value. Lines represent the means of simulations.

Indeed, in the simulation period, the NPP and carbon mass of the montane boreal forest increased. Combined with the analysis of single-factor changes, carbon stocks decreased as temperature rose but increased as CO_2 concentration increased. The actual temperature increased in the study area from 1964 to 2013, precipitation fluctuated, and CO_2 concentration increased, and the positive effect of CO_2 concentration contributed more to the carbon stocks than the negative effect of temperature.

3.3. The Sensitivity Analysis of Carbon Mass of Montane Boreal Forest to Extreme Weather

We analyzed the extremes in weather events from 1964–2013 in stations we chose by ranking the mean value of heat and water in concentration months (annual mean temperature from April to October, and annual mean precipitation from May to September) of each year.

Under warm and wet conditions, NPP increased, and under warm conditions alone, NPP increased more (Figure 7). Under cold and dry conditions, NPP decreased, and it decreased further under dry conditions alone. However, the living biomass was greatly suppressed under dry conditions alone and decreased by 42% compared to the no-change simulation; mean decrease in NPP was 1.816 kg/m² per five years. Under the other three conditions, biomass production performed better than under the actual climate. Biomass production increased (10% at the end of period) most notably under wet conditions, with a mean increase of 1.085 kg/m² per five years.

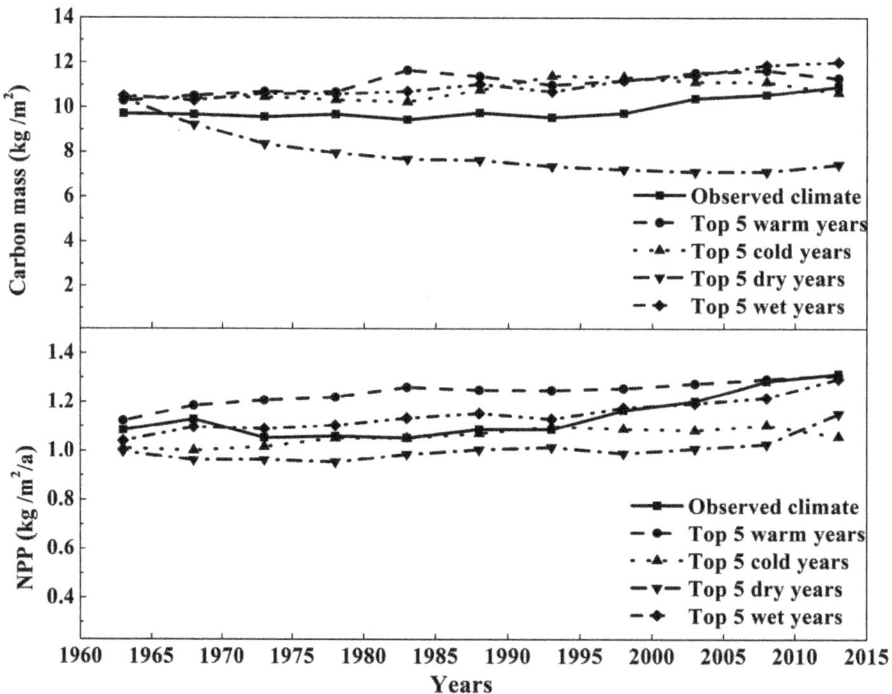

Figure 7. LPJ-GUESS output showing the effects of extreme weather conditions on stored carbon (top panel) and NPP (bottom panel). Local weather data were used to determine the most extreme high /low temperature and high/low rainfall from a 5-year period.

4. Discussion

4.1. Model Applicability and Carbon Mass at Different Altitudes

Our LPJ-GUESS model simulation showed that the main distribution of *Picea crassifolia* in the Qilian Mountains was at 2700 to 3100 m, and the average carbon storage was 10.503 kg/m^2, with a maximum of 11.787 kg/m^2. The biomass pools at elevations 2300–2700 m contained more carbon than those at 3100–3500 m for the different dominant species.

The vertical distribution pattern of *Picea crassifolia* influenced the distribution of carbon density and NPP. Altitude was a dominant factor influencing the soil organic carbon concentration and community pattern of *Picea crassifolia* [52]; further, altitude exerted a strong influence on the growth of *Picea crassifolia* by affecting the microclimate, including air temperature and humidity, and soil moisture [24]. Trees were low and forest density was often low at low altitudes, increasing gradually at mid-altitudes, and exhibiting a pattern of isolated trees or scattered patches at the upper growth limit [53]. The most suitable conditions for the growth of *Picea crassifolia* forest were observed between 2800 and 2900 m [49], with grassland between 3000 to 3700 m [54]. Density and basal area of *Picea crassifolia* were also higher between 2650 to 3100 m than at other altitudes, and beyond 3100 m, the density decreased with an increase in altitude [55].

Evergreen conifers in cold and high-altitude zones of the montane boreal forest have lower carbon biomass due to low photosynthesis and respiration rates than that in warmer habitats [56]. Carbon mass in the biomass pools examined in this study, i.e., 5.462 kg·C/m^2, was similar to the estimate for northern montane boreal forests ranging from 4.2 to 5.3 kg·C/m^2 [47]. The focus of current research in the Qilian Mountains is on the main tree species, *Picea crassifolia*, but results are not uniform. Initial surveys at the Sidalong forest farm on the Qilian Mountain showed that *Picea crassifolia* biomass was about 24.298 kg/m^2 (Chang et al., 1995). Recently, the aboveground carbon stocks of *Picea crassifolia* at the north-eastern edge of Qilian Mountains were calculated as 4.3 kg/C·m^2 [47]. The estimate of the above-ground biomass in southern Qilian Mountains was between 0.1885 and 22.065 kg/m^2 [57]. Our simulation of *Picea crassifolia* NPP was between 0.52–0.58 kg·C/m^2/year, which is in the range of 0.3–0.7 kg·C/m^2/year for boreal ecosystem productivity examined by remote sensing [58], and higher than the mean NPP (0.38 kg·C/m^2/year) of Qilian Mountains [59]. Hydrothermal conditions drove carbon density differences in boreal forests; also, carbon density first increased, and then decreased with stand age, with the highest value at age 183 years [28]. Thus, carbon density calculations were different due to vertical distribution of vegetation patterns and the age of the forest in this study.

The carbon mass and NEE change in *Picea crassifolia* forest ecosystem we calculated revealed that this mountain forest tended to becoming a carbon source from a carbon sink in the last 50 years; this trend was slow and even reversed in the 2000s. The carbon mass of *Picea crassifolia* decreased at low elevations and increased at high elevations, showing an upward trend in vegetation distribution, but the highest carbon mass proportion region for the forest did not change in the study period. Similar to other boreal forests, the mountain boreal forest also tended to become a carbon source under climate change [60–62], but this trend in the Qilian Mountains slowed in the 2000s.

The LPJ-GUESS model has been used in a large number of studies, and its performance has been evaluated several times. The model has also been used in China. Combined LPJ-GUESS and High Accuracy Surface Modeling (HASM) allowed for an economical estimation of forest biomass and the research used climate data from 735 meteorological stations in Chinese mainland from 1950–2010 and compared the results with the Seventh National Forest Resources Inventory data in China [40]. The results presented in this article indicated that the LPJ-GUESS model was suitable for these 735 meteorological stations; stations used in our study were contained in those stations. A carbon balance calculation by LPJ-GUESS of three deciduous forests in a mountainous area near Beijing showed that the model can be applied to a warm temperate forest in China [37]. Also, the forest production and carbon dynamics study of Masson Pine Forest in the Jigongshan region demonstrated that the model can simulate growth dynamics of subtropical forests [39]. Finally, the simulation of the carbon cycle of a *Larix chinensis* forest at Taibai Mountains, China, indicated the model was suitable for analyzing vegetation characteristics in mountainous areas [38]. Similar to previous biomass calculations (Table 2), we found that the LPJ-GUESS model was suitable for simulating carbon change of the montane boreal forest in northwestern China. The carbon density calculated by the model is based only on foliage, fine roots, and sapwood and heartwood; thus, the results of the model can be expected to be lower than results of some field measurements.

4.2. The Effects of The Climate Variables on Carbon Cycling in Montane Boreal Forest

Carbon density and NPP in this montane boreal forest ecosystem were concentrated at 2700–3100 m. Mean carbon mass increased by 13.37%, with a 5-year mean of 1.202 kg/m^2 from 1964 to 2013; mean NPP increased by 0.023 kg/m^2/year per five years.

In this study, NPP clearly increased as temperature increased and under warm conditions. The effect of temperature on carbon storage was complex in this montane boreal forest. Increasing temperature limited the storage of carbon, while decreasing temperature promoted its storage. Under warm conditions, carbon storage increased, and if temperature decreased considerably, carbon mass was reduced. The temperature sensitivity of biomass allocation may be an important but not obvious regulator of the carbon cycle in the boreal forest [19,56]. With an increase in mean annual temperature, and only a modest expected increase in precipitation, a shift in boreal forest may be observed to a woodland/shrubland ecosystem type, which is more suited for such an environment [63]. Further, the overall warming trend may lead to earlier seasonal plant growth; however, incomplete development of green tissues exposed to colder environments may negatively influence tree growth and carbon storage [6]. To the contrary, some researchers concluded that increased temperature and longer growing seasons will increase the NPP and enhance carbon uptake of boreal forests; this response, however, was weak in interannual variations [64]. Boreal forests store vast amounts of carbon in soil, but the temperature effects on soils are challenging to measure. The temperature relationship with above-ground carbon density was affected by baseline temperature such that, at mean annual temperature <8 °C, the relationship was positive, and negative in regions with a mean annual temperature >10 °C for mature boreal forests [65].

Precipitation is positively related to the biomass and NPP of the montane boreal forest. Precipitation has a much greater impact on montane boreal forests than temperature; especially in reduced precipitation and in dry environments, both biomass and NPP decreased significantly and hindered productivity and biomass storage. Drought conditions likely affected the radial growth of *Picea crassifolia* forest in Qilian Mountains in the last half-century [66]. Drought was the major driver of the release of total original carbon, which reduced soil respiration and NPP in a large boreal watershed [67]. Extreme drought decreases carbon assimilation and reduces the carbon sink strength of forests [6]. Under extremely dry conditions, boreal forests may degrade to low-productivity open woodlands [63].

In this study, the effect of CO_2 concentration on the biomass and NPP of montane boreal forest was also consistent, and the parameters were positively correlated. We showed that, if the CO_2 concentration remained at the level present 50 years ago, the ability of the montane boreal forest to produce and store carbon would decrease. However, the capacity of forests to store carbon improved with increasing CO_2 concentration, and the higher the CO_2 amounts, the greater the carbon storage capacity. Although not as pronounced as that in tropical and temperate forests, the increase in CO_2 concentration led to an increase in NPP and carbon stocks in boreal forests both at local and global scales [62,68]. A physiologically-based forest model showed that the total NPP and the carbon storage in biomass improved under elevated CO_2 concentration due to an increase in net photosynthetic rates at leaf-level and smaller shifts in carbon residence time [56,67]. Using combined satellite and ground observations for 1950–2011, Denos and others (2013) demonstrated that the sequestration of atmospheric CO_2 increased due to higher net CO_2 uptake associated with spring and fall growth extensions in northern ecosystems. Increasing atmospheric CO_2 is expected to increase boreal forest carbon in western China [69]. A simulation by the BIOME-BGC model in *Picea crassifolia* forest showed that the effect of CO_2 concentration on NPP was more significant than that of climate change for future climate [70].

5. Conclusions

The carbon mass of montane boreal forest simulated by LPJ-GUESS was comparable, though lower than that reported in other studies. Carbon storage was concentrated at the altitude of 2700–3100 m in the Qilian Mountains. Within this altitudinal range, carbon storage of *Picea crassifolia* increased by 1.202 kg/m^2 per five years, and NPP was between 0.52–0.58 $kg \cdot C/m^2/year$ during the last 50 years. *Picea crassifolia* exhibited a trend toward climbing higher in elevation and becoming a carbon source in the last 50 years.

The boreal forest is significantly affected by climate change and has a slow recovery process. Under steady precipitation, carbon storage increased with increasing atmospheric CO_2 concentration despite the negative effects of warming on montane boreal forest in the Qilian Mountains in the last 50 years. A lack of water is the greatest threat for carbon storage in *Picea crassifolia*. Under the changing CO_2 concentration and precipitation conditions, carbon mass and NPP were positively correlated. The relationship between climate factors, CO_2 concentration, extreme conditions, and carbon storage was largely impacted by stand age, geographical location, altitude, and the environment; therefore, further research needs to examine the drivers of interannual variability in the carbon cycle at different scales and under different conditions of climate change.

Acknowledgments: We are very grateful to Kathryn Piatek for her comments and editorial assistance. This work was supported by the National Natural Science Foundation of China (No. 41621001, 41522102 and 41601051) and the Foundation for Excellent Youth Scholars of "Northwest Institute of Eco-Environment and Resources", Chinese Academy of Sciences.

Author Contributions: Zhibin He and Shu Fang conceived and designed the experiments; Shu Fang performed the experiments, analyzed the data and wrote the paper; Jun Du, Longfei Chen, Pengfei Lin, and Minmin Zhao optimized the experiment and modified the manuscript.

Conflicts of Interest: The authors declare no conflict of interest. The founding sponsors had no role in the design of the study; in the collection, analyses, or interpretation of data; in the writing of the manuscript, and in the decision to publish the results.

References

1. Rössler, O.; Diekkrüger, B.; Löffler, J. Potential drought stress in a Swiss mountain catchment—Ensemble forecasting of high mountain soil moisture reveals a drastic decrease, despite major uncertainties. *Water Resour. Res.* **2012**, *48*. [CrossRef]
2. Hartl-Meier, C.; Dittmar, C.; Zang, C.; Rothe, A. Mountain forest growth response to climate change in the Northern Limestone Alps. *Trees* **2014**, *28*, 819–829. [CrossRef]
3. Savva, Y.; Oleksyn, J.; Reich, P.B.; Tjoelker, M.G.; Vaganov, E.A.; Modrzynski, J. Interannual growth response of Norway spruce to climate along an altitudinal gradient in the Tatra Mountains, Poland. *Trees* **2006**, *20*, 735–746. [CrossRef]
4. Affolter, P.; Büntgen, U.; Esper, J.; Rigling, A.; Weber, P.; Luterbacher, J.; Frank, D. Inner alpine conifer response to 20th century drought swings. *Eur. J. For. Res.* **2010**, *129*, 289–298. [CrossRef]
5. Orlowsky, B.; Seneviratne, S.I. Global changes in extreme events: Regional and seasonal dimension. *Clim. Chang.* **2012**, *110*, 669–696. [CrossRef]
6. Frank, D.; Reichstein, M.; Bahn, M.; Thonicke, K.; Frank, D.; Mahecha, M.D.; Smith, P.; Velde, M.; Vicca, S.; Babst, F. Effects of climate extremes on the terrestrial carbon cycle: Concepts, processes and potential future impacts. *Glob. Chang. Biol.* **2015**, *21*, 2861–2880. [CrossRef] [PubMed]
7. Thurner, M.; Beer, C.; Carvalhais, N.; Forkel, M.; Santoro, M.; Tum, M.; Schmullius, C. Large-scale variation in boreal and temperate forest carbon turnover rate related to climate. *Geophys. Res. Lett.* **2016**, *43*, 4576–4585. [CrossRef]
8. Schlesinger, W.H.; Dietze, M.C.; Jackson, R.B.; Phillips, R.P.; Rhoades, C.C.; Rustad, L.E.; Vose, J.M. Forest biogeochemistry in response to drought. *Glob. Chang. Biol.* **2016**, *22*, 2318–2328. [CrossRef] [PubMed]

9. Stocker, T. *Climate Change 2013: The Physical Science Basis: Working Group I Contribution to the Fifth Assessment Report of the Intergovernmental Panel on Climate Change*; Cambridge University Press: Cambridge, UK, 2014.

10. McCain, C.M.; Colwell, R.K. Assessing the threat to montane biodiversity from discordant shifts in temperature and precipitation in a changing climate. *Ecol. Lett.* **2011**, *14*, 1236–1245. [CrossRef] [PubMed]

11. Allen, C.D.; Macalady, A.K.; Chenchouni, H.; Bachelet, D.; McDowell, N.; Vennetier, M.; Kitzberger, T.; Rigling, A.; Breshears, D.D.; Hogg, E.T. A global overview of drought and heat-induced tree mortality reveals emerging climate change risks for forests. *For. Ecol. Manag.* **2010**, *259*, 660–684. [CrossRef]

12. Kurz, W.A.; Dymond, C.C.; Stinson, G.; Rampley, G.J.; Neilson, E.T.; Carroll, A.L.; Ebata, T.; Safranyik, L. Mountain pine beetle and forest carbon feedback to climate change. *Nature* **2008**, *452*, 987–990. [CrossRef] [PubMed]

13. Ruiz-Labourdette, D.; Nogués-Bravo, D.; Ollero, H.S.; Schmitz, M.F.; Pineda, F.D. Forest composition in mediterranean mountains is projected to shift along the entire elevational gradient under climate change. *J. Biogeogr.* **2012**, *39*, 162–176. [CrossRef]

14. Trumbore, S.; Brando, P.; Hartmann, H. Forest health and global change. *Science* **2015**, *349*, 814–818. [CrossRef] [PubMed]

15. Rocca, M.E.; Brown, P.M.; MacDonald, L.H.; Carrico, C.M. Climate change impacts on fire regimes and key ecosystem services in rocky mountain forests. *For. Ecol. Manag.* **2014**, *327*, 290–305. [CrossRef]

16. Allen, K.A.; Lehsten, V.; Hale, K.; Bradshaw, R. Past and future drivers of an unmanaged carbon sink in European temperate forest. *Ecosystems* **2016**, *19*, 545–554. [CrossRef]

17. Williams, C.A.; Collatz, G.J.; Masek, J.; Huang, C.; Goward, S.N. Impacts of disturbance history on forest carbon stocks and fluxes: Merging satellite disturbance mapping with forest inventory data in a carbon cycle model framework. *Remote Sens. Environ.* **2014**, *151*, 57–71. [CrossRef]

18. Wit, H.A.; Bryn, A.; Hofgaard, A.; Karstensen, J.; Kvalevåg, M.M.; Peters, G.P. Climate warming feedback from mountain birch forest expansion: Reduced albedo dominates carbon uptake. *Glob. Chang. Biol.* **2014**, *20*, 2344–2355. [CrossRef] [PubMed]

19. Elkin, C.; Gutiérrez, A.G.; Leuzinger, S.; Manusch, C.; Temperli, C.; Rasche, L.; Bugmann, H. A 2 °C warmer world is not safe for ecosystem services in the European alps. *Glob. Chang. Biol.* **2013**, *19*, 1827–1840. [CrossRef] [PubMed]

20. He, Z.; Zhao, W.; Liu, H.; Tang, Z. Effect of forest on annual water yield in the mountains of an arid inland river basin: A case study in the pailugou catchment on northwestern China's Qilian Mountains. *Hydrol. Process.* **2012**, *26*, 613–621. [CrossRef]

21. Du, J.; He, Z.; Yang, J.; Chen, L.; Zhu, X. Detecting the effects of climate change on canopy phenology in coniferous forests in semi-arid mountain regions of China. *Int. J. Remote Sens.* **2014**, *35*, 6490–6507. [CrossRef]

22. Baranova, A.; Schickhoff, U.; Wang, S.; Jin, M. Mountain pastures of Qilian Shan: Plant communities, grazing impact and degradation status (Gansu province, NW China). *Hacquetia* **2016**, *15*, 21–35. [CrossRef]

23. Chen, L.; He, Z.; Du, J.; Yang, J.; Zhu, X. Patterns and controls of soil organic carbon and nitrogen in alpine forests of northwestern China. *For. Sci.* **2015**, *61*, 1033–1040. [CrossRef]

24. He, Z.; Zhao, W.; Zhang, L.; Liu, H. Response of tree recruitment to climatic variability in the alpine treeline ecotone of the Qilian Mountains, northwestern China. *For. Sci.* **2013**, *59*, 118–126. [CrossRef]

25. He, Z.; Du, J.; Zhao, W.; Yang, J.; Chen, L.; Zhu, X.; Chang, X.; Liu, H. Assessing temperature sensitivity of subalpine shrub phenology in semi-arid mountain regions of China. *Agric. For. Meteorol.* **2015**, *213*, 45–52. [CrossRef]

26. Chang, X.; Zhao, W.; He, Z. Radial pattern of sap flow and response to microclimate and soil moisture in Qinghai spruce (*Picea crassifolia*) in the upper Heihe river basin of arid northwestern China. *Agric. For. Meteorol.* **2014**, *187*, 14–21. [CrossRef]

27. Tian, F.; Zhao, C.; Feng, Z.-D. Simulating evapotranspiration of Qinghai spruce (*Picea crassifolia*) forest in the Qilian Mountains, northwestern China. *J. Arid Environ.* **2011**, *75*, 648–655. [CrossRef]

28. Peng, S.; Zhao, C.; Zheng, X.; Xu, Z.; He, L. Spatial distribution characteristics of the biomass and carbon storage of Qinghai spruce (*Picea crassifolia*) forests in Qilian mountains. *Chin. J. Appl. Ecol.* **2011**, *22*, 1689–1694.

29. Thomas, S.C.; Martin, A.R. Carbon content of tree tissues: A synthesis. *Forests* **2012**, *3*, 332–352. [CrossRef]

30. Smith, B.; Prentice, I.C.; Sykes, M.T. Representation of vegetation dynamics in the modelling of terrestrial ecosystems: Comparing two contrasting approaches within European climate space. *Glob. Ecol. Biogeogr.* **2001**, *10*, 621–637. [CrossRef]

31. Hickler, T.; Smith, B.; Sykes, M.T.; Davis, M.B.; Sugita, S.; Walker, K. Using a generalized vegetation model to simulate vegetation dynamics in northeastern USA. *Ecology* **2004**, *85*, 519–530. [CrossRef]

32. Ahlström, A.; Schurgers, G.; Arneth, A.; Smith, B. Robustness and uncertainty in terrestrial ecosystem carbon response to CMIP5 climate change projections. *Environ. Res. Lett.* **2012**, *7*, 044008. [CrossRef]

33. Korth, H.; Tsyganenko, N.A.; Johnson, C.L.; Philpott, L.C.; Anderson, B.J.; Al Asad, M.M.; Solomon, S.C.; McNutt, R.L. Modular model for mercury's magnetospheric magnetic field confined within the average observed magnetopause. *J. Geophys. Res. Space Phys.* **2015**, *120*, 4503–4518. [CrossRef] [PubMed]

34. Knorr, W.; Arneth, A.; Jiang, L. Demographic controls of future global fire risk. *Nat. Clim. Chang.* **2016**, *6*, 781–785. [CrossRef]

35. De Kauwe, M.G.; Medlyn, B.E.; Zaehle, S.; Walker, A.P.; Dietze, M.C.; Hickler, T.; Jain, A.K.; Luo, Y.; Parton, W.J.; Prentice, I.C.; et al. Forest water use and water use efficiency at elevated CO_2: A model-data intercomparison at two contrasting temperate forest face sites. *Glob. Chang. Biol.* **2013**, *19*, 1759–1779. [CrossRef] [PubMed]

36. Rosenzweig, C.; Elliott, J.; Deryng, D.; Ruane, A.C.; Müller, C.; Arneth, A.; Boote, K.J.; Folberth, C.; Glotter, M.; Khabarov, N.; et al. Assessing agricultural risks of climate change in the 21st century in a global gridded crop model intercomparison. *Proc. Natl. Acad. Sci. USA* **2014**, *111*, 3268–3273. [CrossRef] [PubMed]

37. Liu, R.G.; Li, N.; Su, H.X.; Sang, W.G. Simulation and analysis on future carbon balance of three deciduous forests in Beijing mountain area, warm temperate zone of China. *Chin. J. Plant Ecol.* **2009**, *33*, 516–534.

38. Li, L.; He, X.; Hu, L.; Li, J. Simulation of the carbon cycle of *Larix chinensis* forest during 1958 and 2008 at Taibai Mountain, China. *Acta Ecol. Sin.* **2013**, *33*, 2845–2855.

39. Peng, X.; Cheng, R.; Xiao, W.; Wang, R.; Wang, X.; Liu, Z. Productivity and carbon dynamic of the masson pine stands in Jigongshan region based on lpj-guess model. *Sci. Silvae Sin.* **2013**, *49*, 7–8.

40. Zhao, M.; Yue, T.; Zhao, N.; Sun, X.; Zhang, X. Combining lpj-guess and hasm to simulate the spatial distribution of forest vegetation carbon stock in China. *J. Geogr. Sci.* **2014**, *24*, 249–268. [CrossRef]

41. Sitch, S. *The Role of Vegetation Dynamics in the Control of Atmospheric CO_2 Content*; Department of Ecology, Lund University: Lund, Sweden, 2000.

42. Sitch, S.; Smith, B.; Prentice, I.C.; Arneth, A.; Bondeau, A.; Cramer, W.; Kaplans, J.O.; Levis, S.; Lucht, W.; Sykes, M.T.; et al. Evaluation of ecosystem dynamica, olant geography and terrestrial carbon cycling in the lpj dynamic global vegetation model. *Glob. Chang. Biol.* **2003**, *9*, 161–185. [CrossRef]

43. Hickler, T.; Fronzek, S.; Araújo, M.B.; Schweiger, O.; Thuiller, W.; Sykes, M.T. An ecosystem model-based estimate of changes in water availability differs from water proxies that are commonly used in species distribution models. *Glob. Ecol. Biogeogr.* **2009**, *18*, 304–313. [CrossRef]

44. Hickler, T.; Vohland, K.; Feehan, J.; Miller, P.A.; Smith, B.; Costa, L.; Giesecke, T.; Fronzek, S.; Carter, T.R.; Cramer, W. Projecting the future distribution of European potential natural vegetation zones with a generalized, tree species-based dynamic vegetation model. *Glob. Ecol. Biogeogr.* **2012**, *21*, 50–63. [CrossRef]

45. Wu, Q.; Hu, Q.W.; Zheng, L.; Zhang, F.; Song, M.H.; Liu, X.D. Variations of leaf lifespan and leaf mass per area of *Picea crassifolia* along altitude gradient. *Acta Bot. Boreal. Occident. Sin.* **2010**, *30*, 1689–1694.

46. Chen, F.; Yuan, Y.-J.; Wei, W.-S.; Yu, S.-L.; Fan, Z.-A.; Zhang, R.-B.; Zhang, T.-W.; Li, Q.; Shang, H.-M. Temperature reconstruction from tree-ring maximum latewood density of Qinghai spruce in middle Hexi corridor, China. *Theor. Appl. Clim.* **2012**, *107*, 633–643. [CrossRef]

47. Wagner, B.; Liang, E.; Li, X.; Dulamsuren, C.; Leuschner, C.; Hauck, M. Carbon pools of semi-arid *Picea crassifolia* forests in the Qilian Mountains (north-eastern Tibetan Plateau). *For. Ecol. Manag.* **2015**, *343*, 136–143. [CrossRef]

48. Jing, W.-M.; Liu, X.-D.; Zhao, W.-J.; Ma, Y. Study on biomass and net productivity of typical forest stand in the Qilian mountains. *J. Gansu Agric. Univ.* **2011**, *6*, 017.

49. Zhang, L.; Yu, P.; Wang, Y.; Wang, S.; Liu, X. Biomass change of middle aged forest of Qinghai spruce along an altitudinal gradient on the north slope of Qilian mountains. *Sci. Silvae Sin.* **2015**, *51*, 1–7.

50. Allen, D.; Mackie, D.; Wei, M. Groundwater and climate change: A sensitivity analysis for the grand forks aquifer, southern British Columbia, Canada. *Hydrogeol. J.* **2004**, *12*, 270–290. [CrossRef]

51. Katz, R. Extreme value theory for precipitation: Sensitivity analysis for climate change. *Adv. Water Res.* **1999**, *23*, 133–139. [CrossRef]

52. Zhao, C.; Chen, L.; Ma, F.; Yao, B.; Liu, J. Altitudinal differences in the leaf fitness of juvenile and mature alpine spruce trees (*Picea crassifolia*). *Tree Phys.* **2008**, *28*, 133–141. [CrossRef]

53. Zhao, C.; Nan, Z.; Cheng, G.; Zhang, J.; Feng, Z. Gis-assisted modelling of the spatial distribution of Qinghai spruce (*Picea crassifolia*) in the Qilian mountains, northwestern china based on biophysical parameters. *Ecol. Model.* **2006**, *191*, 487–500. [CrossRef]

54. Liang, B.; Di, L.; Zhao, C.; Peng, S.; Peng, H.; Wang, C.; Wang, Y.; Liu, Y. Altitude distribution of aboveground biomass of typical shrubs in the tianlaochi watershed of Qilian mountains. *Acta Agrestia Sin.* **2013**, *21*, 664–669.

55. Zhang, L.; Liu, H. Response of *Picea crassifolia* population to climate change at the treeline ecotones in Qilian mountains. *Sci. Silvae Sin.* **2012**, *1*, 006.

56. Reich, P.B.; Rich, R.L.; Lu, X.; Wang, Y.-P.; Oleksyn, J. Biogeographic variation in evergreen conifer needle longevity and impacts on boreal forest carbon cycle projections. *Proc. Natl. Acad. Sci. USA* **2014**, *111*, 13703–13708. [CrossRef] [PubMed]

57. Tian, X.; Li, Z.; Su, Z.; Chen, E.; Tol, C.V.D.; Li, X.; Guo, Y.; Li, L.; Ling, F. Estimating montane forest above-ground biomass in the upper reaches of the Heihe river basin using landsat-tm data. *Int. J. Remote Sens.* **2014**, *35*, 7339–7362. [CrossRef]

58. Zhou, Y.; Zhu, Q.; Chen, J.M.; Wang, Y.Q.; Liu, J.; Sun, R.; Tang, S. Observation and simulation of net primary productivity in Qilian mountain, western china. *J. Environ. Manag.* **2007**, *85*, 574–584. [CrossRef] [PubMed]

59. Wang, P.; Sun, R.; Hu, J.; Zhu, Q.; Zhou, Y.; Li, L.; Chen, J.M. Measurements and simulation of forest leaf area index and net primary productivity in northern China. *J. Environ. Manag.* **2007**, *85*, 607–615. [CrossRef] [PubMed]

60. Wramneby, A.; Smith, B.; Zaehle, S.; Sykes, M.T. Parameter uncertainties in the modelling of vegetation dynamics—Effects on tree community structure and ecosystem functioning in European forest biomes. *Ecol. Model.* **2008**, *216*, 277–290. [CrossRef]

61. Smith, B.; Knorr, W.; Widlowski, J.-L.; Pinty, B.; Gobron, N. Combining remote sensing data with process modelling to monitor boreal conifer forest carbon balances. *For. Ecol. Manag.* **2008**, *255*, 3985–3994. [CrossRef]

62. Hickler, T.; Smith, B.; Prentice, I.C.; Mjöfors, K.; Miller, P.; Arneth, A.; Sykes, M.T. CO_2 fertilization in temperate face experiments not representative of boreal and tropical forests. *Glob. Chang. Biol.* **2008**, *14*, 1531–1542. [CrossRef]

63. Clemmensen, K.; Bahr, A.; Ovaskainen, O.; Dahlberg, A.; Ekblad, A.; Wallander, H.; Stenlid, J.; Finlay, R.; Wardle, D.; Lindahl, B. Roots and associated fungi drive long-term carbon sequestration in boreal forest. *Science* **2013**, *339*, 1615–1618. [CrossRef] [PubMed]

64. Piao, S.; Liu, Z.; Wang, T.; Peng, S.; Ciais, P.; Huang, M.; Ahlstrom, A.; Burkhart, J.F.; Chevallier, F.; Janssens, I.A. Weakening temperature control on the interannual variations of spring carbon uptake across northern lands. *Nat. Clim. Chang.* **2017**, *7*, 359–363. [CrossRef]

65. Liu, Y.; Yu, G.; Wang, Q.; Zhang, Y. How temperature, precipitation and stand age control the biomass carbon density of global mature forests. *Glob. Ecol. Biogeogr.* **2014**, *23*, 323–333. [CrossRef]

66. Zhang, Y.; Shao, X.; Wilmking, M. Dynamic relationships between *Picea crassifolia* growth and climate at upper treeline in the Qilian Mts., northeast Tibetan Plateau, China. *Dendrochronologia* **2011**, *29*, 185–199. [CrossRef]

67. Lepistö, A.; Futter, M.N.; Kortelainen, P. Almost 50 years of monitoring shows that climate, not forestry, controls long-term organic carbon fluxes in a large boreal watershed. *Glob. Chang. Boil.* **2014**, *20*, 1225–1237. [CrossRef] [PubMed]

68. Schimel, D.; Stephens, B.B.; Fisher, J.B. Effect of increasing CO_2 on the terrestrial carbon cycle. *Proc. Natl. Acad. Sci. USA* **2015**, *112*, 436–441. [CrossRef] [PubMed]

69. Friend, A.D.; Lucht, W.; Rademacher, T.T.; Keribin, R.; Betts, R.; Cadule, P.; Ciais, P.; Clark, D.B.; Dankers, R.; Falloon, P.D. Carbon residence time dominates uncertainty in terrestrial vegetation responses to future climate and atmospheric CO_2. *Proc. Natl. Acad. Sci. USA* **2014**, *111*, 3280–3285. [CrossRef] [PubMed]

70. Peng, S.; Zhao, C.; Chen, Y.; Xu, Z. Simulating the productivity of a subalpine forest at high elevations under representative concentration pathway (RCP) scenarios in the Qilian Mountains of northwest China. *Scand. J. For. Res.* **2016**, *32*, 1–31.

Article

Forest Floor and Mineral Soil Respiration Rates in a Northern Minnesota Red Pine Chronosequence

Matthew Powers [1,*], Randall Kolka [2], John Bradford [3], Brian Palik [2] and Martin Jurgensen [4]

[1] Department of Forest Engineering Resources and Management, Oregon State University,
    280 Peavy Hall, Corvallis, OR 97331, USA
[2] USDA Forest Service Northern Research Station, 1831 Hwy 169 E, Grand Rapids, MN 55744, USA;
    rkolka@fs.fed.us (R.K.); bpalik@fs.fed.us (B.P.)
[3] US Geological Survey Southwest Biological Science Center, 2255 N. Gemini Dr., Flagstaff, AZ 86001, USA;
    jbradford@usgs.gov
[4] School of Forest Resources and Environmental Science, Michigan Technological University,
    1400 Townsend Drive, Houghton, MI 49931, USA; mfjurgen@mtu.edu
* Correspondence: matthew.powers@oregonstate.edu; Tel.: +1-541-737-6550

Received: 31 October 2017; Accepted: 25 December 2017; Published: 29 December 2017

Abstract: We measured total soil CO_2 efflux (R_S) and efflux from the forest floor layers (R_{FF}) in red pine (*Pinus resinosa* Ait.) stands of different ages to examine relationships between stand age and belowground C cycling. Soil temperature and R_S were often lower in a 31-year-old stand (Y31) than in 9-year-old (Y9), 61-year-old (Y61), or 123-year-old (Y123) stands. This pattern was most apparent during warm summer months, but there were no consistent differences in R_{FF} among different-aged stands. R_{FF} represented an average of 4–13% of total soil respiration, and forest floor removal increased moisture content in the mineral soil. We found no evidence of an age effect on the temperature sensitivity of R_S, but respiration rates in Y61 and Y123 were less sensitive to low soil moisture than R_S in Y9 and Y31. Our results suggest that soil respiration's sensitivity to soil moisture may change more over the course of stand development than its sensitivity to soil temperature in red pine, and that management activities that alter landscape-scale age distributions in red pine forests could have significant impacts on rates of soil CO_2 efflux from this forest type.

Keywords: carbon cycling; *Pinus resinosa*; soil respiration; stand age

1. Introduction

Soil respiration represents about 70% of ecosystem respiration in temperate forests [1,2], and includes a combination of respiration from plant roots, mycorrhizae, and microorganisms in the leaf litter, humus, and mineral soil. Variables such as soil temperature [3,4], soil moisture [2,5–7], litter quality and quantity [4,8], and local stand structure [3,9] exert strong controls over soil respiration in forests. Soil temperature and soil moisture may also covary seasonally, or show varying relationships across sites, leading to confounded effects on soil respiration [10,11]. Changes in these variables that occur over the course of stand development could also lead to changes in soil respiration as forests age. Soil temperatures and moisture availability, for instance, may vary between young, open-canopied regenerating stands and stands with a dense, closed canopy representative of the stem exclusion phase of development. As land management agencies begin to incorporate C storage and sequestration into their management goals there is an increasing need to understand how developmental changes influence belowground C cycling.

Soil C cycling can display a large amount of within-system and among-system variability as forests age and develop. Estimates of soil respiration derived from relationships between net primary production and net ecosystem production suggest a characteristic age-related trend in which the heterotrophic components of soil respiration (micro- and macro-fauna) in temperate forests are expected to decline with age, and the autotrophic portion of soil respiration (plant roots and associated mycorrhizae) are expected to peak in middle-aged stands as net primary production (NPP) and, therefore, substrate supplies peak [12]. Direct measurements of soil respiration in stands of different ages, however, have shown that total soil CO_2 efflux (R_S) can (1) initially increase with age, peaking in young-intermediate-aged, closed canopy stands before declining with age in mature to older stands [4,13,14]; (2) decrease with age [15]; (3) increase steadily from young to old-growth stands [6,16]; or (4) show opposing age-related trends across the geographic range of the same forest type [17]. This variability in age-related soil respiration trends among different systems underscores the need for ecosystem-specific studies.

While respiration from roots and associated mycorrhizae make up the largest proportion of soil respiration [18], estimates of respiration associated with decomposition of the forest floor layers range from 5–48% of total soil respiration in temperate forests [4,18–22]. There are no clearly established patterns of respiration from the forest floor layers (R_{FF}) across stands of different ages, but R_S generally has positive correlations with forest floor mass or thickness [3,15,23]. This suggests that R_{FF} could increase with stand age in forests that are characterized by slow litter decay rates and a resulting increase in forest floor thickness over time. Forests dominated by evergreen conifers, for instance, often accumulate forest floor mass and thickness as they age [24,25], which could lead to increased R_{FF} in older stands. The accumulation of forest floor layers during stand development can also exert a large indirect influence on mineral soil respiration by altering soil temperature and moisture or providing leachate of labile substrate into the mineral soil [26,27].

Understanding how stand age influences C cycling is particularly important because forest management activities have direct impacts on landscape-scale age distributions. Different rotation lengths, for instance, can result in dramatically different abundances of young vs. old stands. This would have significant impacts on soil C losses for systems that show a high degree of variability in soil respiration among age classes. The collection of harvest residues and residual wood as a feedstock for biofuel production can also reduce forest floor and mineral soil C and nutrient stocks [28–30], and litter removal clearly impacts both soil respiration and other soil processes [20,31].

We conducted an experiment to characterize age-related differences in total soil respiration (R_S) and respiration from the forest floor layers (R_{FF}) in red pine (*Pinus resinosa* Ait.) stands aged 9–123 years. Productivity in managed red pine systems peaks between 130 and 140 years across a range of basal areas [32], so we expected R_S to increase steadily with age across our chronosequence due to increases in autotrophic respiration linked to increasing NPP [33]. We also predicted that R_{FF} would increase with age across the chronosequence because forest floor C (and thus, forest floor mass) increases steadily with age up to at least 150 years in red pine forests [25], and forest floor mass is positively correlated with R_S [3,15,23].

2. Materials and Methods

2.1. Study Area

Our study sites included four red pine stands on the Chippewa National Forest in northern Minnesota, USA. The study sites included an open, 9-year-old stand of red pine saplings that were planted following the clearcutting of the previous red pine stand (Y9), a 31-year-old plantation in the early stages of stem exclusion (Y31), a 61-year-old plantation with a well-developed shrub layer and some tree recruitment in the understory (Y61), and a 123 year-old naturally-regenerated stand with a well-develop shrub and understory tree layer (Y123). Three of the stands were located in close proximity (within 1000 m of one another), and the fourth (Y61) was located 5 km to the west.

Forest floor and soil characteristics, live tree density, and basal area varied somewhat by stand age (Table 1). Red pine represented 94–98% of the basal area of stems $\geq$2.5 cm in diameter in Y31, Y61, and Y123, with eastern white pine (*Pinus strobus* L.), paper birch (*Betula papyrifera* Marsh.), balsam fir

(*Abies balsamea* L.), and northern red oak (*Quercus rubra* L.) each representing anywhere from 0–3.4% of the remaining basal area within these stands. Y9 had very little basal area of stems $\geq$2.5 cm in diameter. Hardwoods represented < 2% of basal area in any stand. Beaked hazel (*Corylus cornuta* Marsch.), balsam fir, juneberries (*Amelanchier* spp.), and blueberries (*Vaccinium* spp.) were the dominant understory shrubs and saplings, with a much more developed shrub, sapling, and herbaceous layer in the young, open-canopied Y9 than the other stands.

Study sites were selected to minimize variability in edaphic and physiographic characteristics that could contribute to differences in the soil environment or local microclimate. All stands were located on coarse-textured, excessively drained outwash sands classified as Menahga mixed, frigid Typic Udipsamments to minimize variability in physical soil characteristics that could contribute to differences in soil respiration [34]. Topographic variation across the study sites was very limited with less than 10 m variability in elevation among sites, slopes less than 10%, and plots within each site located on predominantly south or southeast facing aspects. The study area has an average annual temperature of 3.9 °C and average annual precipitation of 700 mm.

Table 1. Forest floor, soil (0–30 cm), and overstory structure characteristics of 9, 31, 61, and 123 year-old red pine stands used to study age-related changes in soil respiration, forest floor respiration, and litter decomposition.

Stand	Age in 2009 (years)	FF Thickness (cm)	FF Mass (kg/m^2)	FF %C	FF %N	FF C:N	Soil %C	Soil %N	Soil C:N	pH A	pH A-30	Overstory Density (TPH)	Basal Area (m^2 ha^{-1})
Y9	9	1.67	1.11	40.23	1.30	30.76	2.09	0.10	19.17	4.03	4.96	67	0.04
Y31	31	4.04	2.59	36.51	0.91	39.94	0.69	0.04	19.92	4.48	4.38	2167	40.06
Y61	61	4.90	2.18	49.38	1.11	44.87	0.89	0.06	15.00	4.51	4.80	633	38.18
Y123	123	4.33	2.43	49.76	1.29	38.81	0.72	0.04	19.51	4.40	4.60	650	40.63

FF = forest floor; TPH = trees ha^{-1}; pH A = soil pH from the surface to the top of the A horizon; pH A-30 = soil pH from the top of the A horizon to a depth of 30 cm; all pH values measured in a $CaCl_2$ solution. Forest floor and soil %C and %N measured on a LECO Total Elemental Analyzer. Note that %C and %N data were not ash-corrected. Forest floor mass is an estimate based on a companion study in the same stands [35] and mass from the actual collars in this study.

2.2. Field and Laboratory Methods

We selected a chronosequence of 4 red pine stands at 9, 31, 61, and 123 years of age. We placed 25 cm (inside diameter) PVC (polyvinyl chloride) collars into the soil in November of 2008 after leaf fall of deciduous trees and shrubs (approximately 6 months prior to our first sampling). Nine collars were placed at three randomly located sample plots in each stand, for a total of 27 collars per stand. The nine collars in each plot were placed in groups of three at a distance of four meters from plot center along transects centered at azimuths of 60, 180, and 300 degrees at each plot.

Since the presence of leaf litter can exert a large indirect influence on mineral soil respiration by altering soil temperature and moisture or providing leachate of labile substrate into the mineral soil [26,27], we used forest floor manipulation treatments to examine direct and indirect impacts of forest floor removal on R_S and estimates of R_{FF}. In this context, "direct" implies the additive contribution of CO_2 efflux from litter decomposition to total R_S, while "indirect" implies the effects that modification of the soil environment due to the presence of an intact litter layer could have on both autotrophic and heterotrophic contributions to R_S from the mineral soil. In April of 2010, we removed the forest floor (down to mineral soil) from one randomly selected collar at each transect (the "no litter" treatment) to evaluate how complete forest floor removal impacts soil respiration. Since complete forest floor removal could create indirect impacts on soil respiration by altering the soil environment, we also removed the forest floor from a second PVC collar at each transect, and replaced it with a removable, 20-mesh (841 micron) bag (filled with litter collected from an area near each transect of collars equal in size to the cross-sectional area of the collars themselves. In this treatment (the "removable bag" treatment) the mesh bags were removed immediately prior to taking soil respiration measurements, with the intention of maintaining mineral soil conditions similar to those under intact litter while removing the direct contribution of CO_2 efflux from the forest floor from estimates of soil respiration. The third collar in each transect was left as an untreated control.

Total soil CO_2 efflux (R_S) was measured on a monthly basis on all 27 collars in each stand from May through October in 2009 (pre-manipulation; i.e., prior to litter removal and removable bag installation) and from April through October in 2010 (post-manipulation; i.e., after litter removal and removable bag installation) using a custom-built, closed-chamber system with an LI-820 gas analyzer (LI-Cor, Inc., Lincoln, NE, USA) connected to a datalogger programmed to record the CO_2 concentrations in the system at 2 s intervals. Equipment malfunctions prevented us from completing the August 2009 measurement.

Since our closed-chamber system calculates CO_2 flux rates by first scrubbing CO_2 out of the chamber and then measuring the rate of increase as chamber CO_2 rises back to a pre-set upper boundary, measurement times on individual collars varied based on instantaneous flux rates. R_S measurements were conducted over a 2–3 day period each month, with all collars in each stand being sampled on a single day in each month. Individual stands were sampled at alternating the times of day (0930–1200 or 1200–1500) from one month to the next in an effort to average out variability associated with taking measurements at different locations at different times of day.

Temperature readings from the top 10 cm of soil suggest soils were frozen from mid-late November until mid-April during both years of measurement, so we assume that soil respiration was minimal during the November to April period. All fresh litter inputs were removed from the no litter treatments on a bi-weekly to monthly basis throughout our 2010 measurements, and respiration measurements in the removable bag treatment were made five to ten minutes after removing the bags from the collars to allow the CO_2 environment within the collar to equilibrate after removing the bag. We measured mineral soil temperature at a depth of 10 cm next to each collar, and measured volumetric soil moisture content in the top 12 cm of the mineral soilusing a HydroSense Soil Water Measurement System consisting of a CS620 sensor attached to a CD620 display (Campbell Scientific, Inc., Logan, UT, USA). Soil moisture measurements were taken at four points around each collar during each soil respiration measurement. After litter treatment in 2010, temperature and moisture measurements for the no litter and removable bag treatments were made within similar treatment areas next to each treatment collar to capture treatment effects without disturbing the soil environment within the collars used for respiration measurements.

We estimated R_{FF} based on the differences of R_S measurements between the control and treatment collars as Equations (1) and (2):

$$R_{FF\ NO\ LITTER} = R_{S\ CONTROL} - R_{S\ NO\ LITTER} \tag{1}$$

and

$$R_{FF\ REMOVABLE\ BAG} = R_{S\ CONTROL} - R_{S\ REMOVABLE\ BAG} \tag{2}$$

where R_{FF} is respiration from the forest floor layers, $R_{S\ CONTROL}$ is the measured respiration from the control collar in a transect, $R_{S\ CONTROL}$ is the measured respiration from the "no litter" treatment collar in that transect (i.e., collars that received a complete litter removal treatment), and $R_{S\ REMOVABLE\ BAG}$ is the measured respiration from the "removable bag" collar in that same transect (i.e., collars in which the litter was removed from the collar and replaced with a mesh bag filled with litter that could be left in place between measurements and removed immediately prior to measurements).

2.3. Analytical Methods

We analyzed litter treatment and stand age effects on R_S, R_{FF}, soil temperature, and soil moisture using linear mixed models. Models included litter treatment, stand age, month of measurement, and the interactions among these three factors as fixed effects along with random effects to account for the spatial nesting of litter treatments within plots and the temporal nesting of measurements within plots. We used averages of litter treatments or litter types from the three transects at each plot as the dependent variable in each analysis. Respiration data, temperature data, and moisture data from 2009 and 2010 were analyzed separately because the measurements spanned different months in the two years. Bonferroni-adjusted t-tests were used to evaluate significant differences between treatment levels. Model assumptions were evaluated with residual plots.

We analyzed relationships between R_S and soil temperature based on the exponential growth function (Equation (3)):

$$R_S = \beta_1 \exp(\beta_2 \times T) \tag{3}$$

where R_S is soil respiration and T is temperature. To include nested random effects (which do not have built-in support for mixed-effects nonlinear models in SAS version 9.2 (SAS Institute, Cary, NC, USA)) to account for repeated measurements on collars representing different treatments within plots, we log-transformed the exponential growth function, and analyzed the soil respiration—temperature relationship using linear mixed models of the form (Equation (4)):

$$LN(R_S) = LN(\beta_1) + (\beta_2) \times T \tag{4}$$

where LN refers to the natural logarithm, R_S is soil respiration, and T is soil temperature. To test for potential stand age and litter treatment effects on the relationship between soil respiration and temperature, we fit models that allowed β_1 and β_2 to vary both individually and in tandem across the levels of both stand age and litter treatment. We used AIC scores to choose the model with the best fit to our data, but also considered the *p*-values of individual model terms to judge the significance of age and litter treatment effects. All models relating R_S to temperature included random effects to account for the use of repeated measurements and the nesting of treatments within plots in each stand. We estimated the sensitivity of soil respiration to temperature using Q_{10} values. We calculated Q10 as Equation (5):

$$Q_{10} = (R_2/R_1)^{(10/(T_2 - T_1))} \tag{5}$$

where R_1 and R_2 are modeled soil respiration rates at temperatures T_1 and T_2, respectively.

We also used linear mixed models to analyze relationships between soil respiration and soil moisture. All soil moisture models followed the general form (Equation (6)):

$$R = \beta_1 + \beta_2 \times SM \tag{6}$$

where R is either an absolute, or one of two normalized measure of soil respiration and SM is volumetric soil moisture. Thus, we ran three sets of models relating soil respiration to soil moisture including (1) one set including our direct, field-based R_S estimates as the dependent variable; (2) a second set using a normalized expression of soil respiration (we refer to this as R_{SN}) calculated as the ratio of observed R_S to the value predicted by our best-fitting temperature model based on soil temperatures at the time of field R_S sampling; and (3) a third set expressing soil respiration as the difference between our observed value from field measurements of R_S, and the value predicted by our best-fitting temperature model based on soil temperatures at the time of field R_S sampling (we refer to this variable as R_{SDIF}).

Like our respiration-temperature models, we fit respiration-moisture models that allowed β_1 and β_2 to vary by both individually and in combination across the levels of stand age and litter type, then used AIC scores to choose the best model. We used residual plots to analyze model assumptions regarding the normality and homogeneity of error variances for all of our regression models, and applied transformations when necessary. We used Bonferroni-adjusted *t*-tests for all multiple comparisons. We calculated a conditional R^2-type goodness of fit value for the respiration—temperature and respiration—moisture models as Equation (7):

$$R^2{}_C = 1 - \left[\Sigma(y_{ij} - \hat{y}_{ij})^2 / \Sigma(y_{ij} - \bar{y})^2 \right] \tag{7}$$

where y_{ij} and $\hat{y}_{ij}$ are the observed and predicted respiration rates for each individual subject and $\bar{y}$ is the overall mean of the observed respiration rates [36]. All statistical analyses were performed using SAS version 9.2 (SAS Institute, Cary, NC, USA) at $\alpha = 0.05$ significance level.

3. Results

3.1. Total Soil Respiration

Litter treatment had significant impacts on R_S ($p = 0.010$ for post-manipulation measurements), and stand age ($p < 0.001$ and $p = 0.001$), month of measurement ($p < 0.001$ for both years), and the stand age x month of measurement interaction ($p < 0.001$ for both years) were also significant in the pre-manipulation and post-manipulation measurement years. There were no significant interactions involving litter treatment in either year of measurement. There were no significant differences in R_S among collars assigned different litter treatments prior to litter treatment ($p = 0.285$), but R_S was significantly lower for the no litter treatment than for the control treatment after litter treatment (Table 2).

Table 2. Effects of forest floor removal treatments on total soil respiration (R_S), forest floor respiration (R_{FF}) (calculated after litter treatment in 2010 only), and soil moisture (SM) before (2009) and after (2010) experimental litter removal treatment in red pine stands. Different letters indicate significant differences among treatments within a column. Respiration rates reported in μmol CO_2 m^{-2} s^{-1}.

Treatment	R_S 2009	R_S 2010	R_{FF} 2010	SM (%) 2009	SM (%) 2010
No Litter	3.23 (a)	2.93 (b)	0.35 (a)	10.36 (a)	14.02 (b)
Removable Bag	3.30 (a)	3.21 (ab)	0.06 (b)	10.62 (a)	12.67 (b)
Control	3.17 (a)	3.27 (a)	–	10.60 (a)	12.79 (a)

Averaged across months, R_S was highest in Y9 and lowest in Y31 in 2009, and generally lowest in Y31, but similar among other stands in 2010 (Table 3). However, the age effect was variable across the individual months of measurement (Figure 1). From May through September of 2009 (Figure 1e) and June through August of 2010 (Figure 1f), R_S was generally lowest in Y31. There were no differences in R_S among stands in October of 2009, or in April, May, or October of 2010, when soil temperatures were at their lowest for each year. Y61 had higher R_S than any other stand in September of 2010, but was generally similar to the Y9 and Y123 stands in other months. Monthly variability in R_S generally paralleled soil temperature trends in both years. In 2009, R_S monthly patterns of Y61 and Y123 closely followed the temperature trend, although R_S in Y9 and Y31 began declining in July while soil temperatures remained high. In 2010, R_S patterns closely followed the monthly temperature trends, showing a steady increase through August followed by a decline in September and October.

Table 3. Mean values of total soil respiration (R_S), volumetric soil moisture (SM), soil temperature (T), and forest floor respiration (R_{FF}) before experimental litter removal (2009) and after litter removal (2010) in red pine stands aged 9 (Y9), 31 (Y31), 61 (Y61), and 123 (Y123) years. Values represent annualized averages across multiple months of measurement in each year. R_{FF} was calculated only after litter removal (i.e., only in 2010). Different letters indicate significant differences among treatments within a column. Respiration rates reported in μmol CO_2 m^{-2} s^{-1}.

Stand ID	R_S 2009	R_S 2010	R_{FF} 2010	T (°C) 2009	T (°C) 2010	SM (%) 2009	SM (%) 2010
Y9	3.78 (a)	3.11 (ab)	0.08 (a)	12.51 (a)	10.02 (a)	13.36 (a)	15.72 (a)
Y31	2.32 (b)	2.46 (b)	0.42 (a)	10.05 (c)	8.98 (b)	9.39 (b)	12.71 (b)
Y61	3.54 (c)	3.76 (a)	0.05 (a)	11.26 (b)	9.99 (a)	9.71 (b)	11.14 (b)
Y123	3.30 (c)	3.20 (ab)	0.27 (a)	10.49 (c)	8.95 (b)	9.96 (b)	13.07 (b)

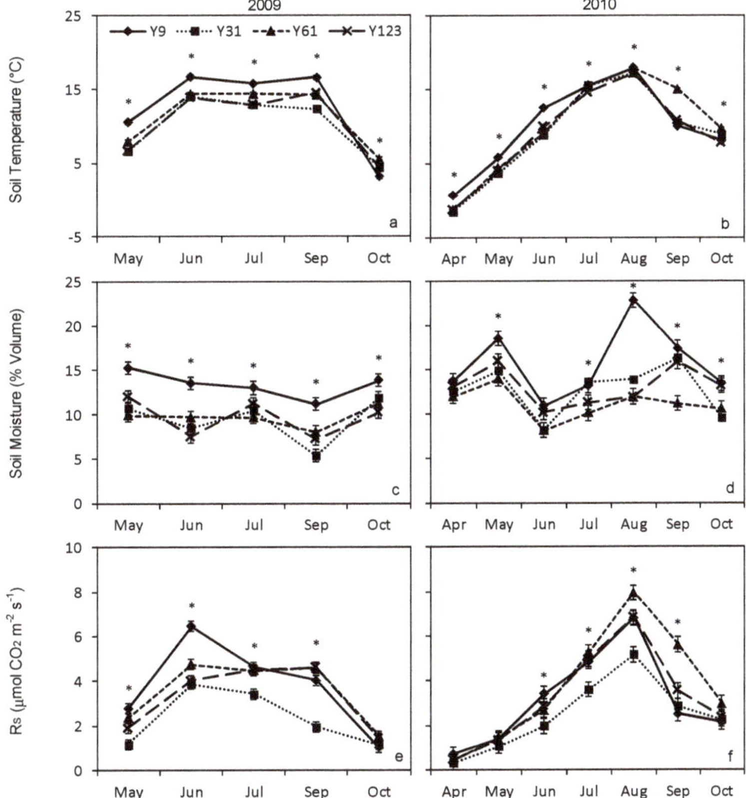

Figure 1. Soil temperatures (a, b), soil moisture (c, d), and soil respiration (R_S, e, f) from two years of measurement in 9, 31, 61, and 123 year-old red pine stands in northern Minnesota, USA. Error bars indicate standard error. An asterisk indicates a month with significant differences among stand ages.

3.2. Forest Floor Respiration

Our estimates of R_{FF} were more than five times higher for the no litter treatment than for the removable bag treatment ($p = 0.044$, Table 2). R_{FF} was not significantly affected by stand age (Table 3) or month of measurement, and there were no significant interactions between stand age, month of measurement, and litter treatment type (i.e., $R_{FF\ LITTER}$ vs. $R_{FF\ REMOVABLE\ BAG}$). R_{FF} represented 13.1% of R_S in the no litter treatment compared to 3.9% of R_S in the removable bag treatment ($p = 0.018$, Table 2). This percentage varied by month of measurement ($p < 0.001$), but the month effect was not consistent across stands of different ages ($p = 0.027$). R_{FF} generally represented the smallest percentage of R_S in April when soil temperatures were lowest across all stands, but differences among other months were highly variable across stands (Figure 2). There were no significant interactions involving litter treatment, so monthly values displayed in Figure 2 have been expressed as the mean of R_{FF} estimates calculated using the removable bag treatment and the mineral soil treatment.

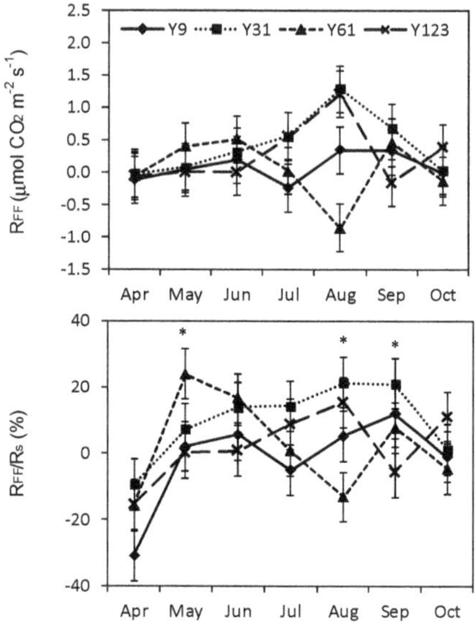

Figure 2. CO_2 efflux from forest floor layers (R_{FF}, top panel) and the percentage of total soil respiration (R_S) represented by the forest floor (bottom panel) in 9, 31, 61, and 123 year-old red pine stands in northern Minnesota, USA. Error bars indicate standard error. An asterisk indicates a month with significant differences among stand ages. R_{FF} estimates displayed here reflect the mean of values calculated from two separate treatments (complete litter removal and temporary removal of litter in a mesh bag).

3.3. Soil Temperature and Moisture

Litter treatment did not have a significant effect on soil temperature in either pre- or post-treatment measurements ($p = 0.999$ and 0.372, respectively), but stand age ($p < 0.001$ for both years) and month of measurement ($p < 0.001$ for both years) each did, and the stand age effect varied across months ($p < 0.001$ for both years). There were no significant interactions involving litter treatment in either year of measurement. Averaged across months, soil temperatures were generally highest in Y9, lowest in Y31 and Y123, and intermediate to high in Y61 (Table 3), but these patterns changed in October, when soil temperatures were lowest in Y9 and highest in Y61 (Figure 1). The temporal trends of soil temperature were somewhat different from 2009 to 2010. In 2009, soil temperatures increased from May to June, were similar in June, July, and August, and then declined rapidly (Figure 1a). In 2010, soil temperatures rose steadily from April through August then declined in September and October (Figure 1b). Soil temperatures in the spring of 2009 were somewhat higher than those during the spring of 2010 but the reverse was true during late summer and autumn.

Litter treatment had a significant impact on soil moisture in 2010 ($p < 0.001$), and stand age, month of measurement, and the stand age x month of measurement interaction were each significant in both the pre-manipulation and post-manipulation years of measurement ($p < 0.001$ for all). There were no significant interactions involving litter treatment in either year of measurement. Although there were no differences between collars assigned different litter treatments prior to manipulation in 2009 ($p = 0.174$), soil moisture was higher in the no litter treatment than in the removable bag and control treatments after litter treatment in 2010 (Table 2). In 2009, soil moisture was typically highest in Y9, and similar among other treatments (Table 3, Figure 1c). In 2010, soil moisture was similar among stands during April and October, highest in Y9 during May and August, and generally low in Y61 during July and September (Figure 1d).

3.4. Relationships between Soil Respiration, Temperature, and Moisture

There was a significant, exponential relationship between R_S and temperature ($p < 0.001$, $R^2_C = 0.787$), but stand age and litter treatment did not have significant effects on the R_S—temperature relationship (Figure 3). There was also a significant, logarithmic relationship between R_S and soil moisture ($p < 0.001$), but model predictions were not well correlated with field observations ($R^2_C = 0.083$). Additionally, the shape of the R_S—moisture relationship varied with stand age ($p < 0.001$ for both soil moisture and the soil moisture x stand age interaction). Our model predicted positive relationships between R_S and moisture in Y9 and Y31, but negative relationships between these two variables in Y61 and Y123 stands (Figure 4).

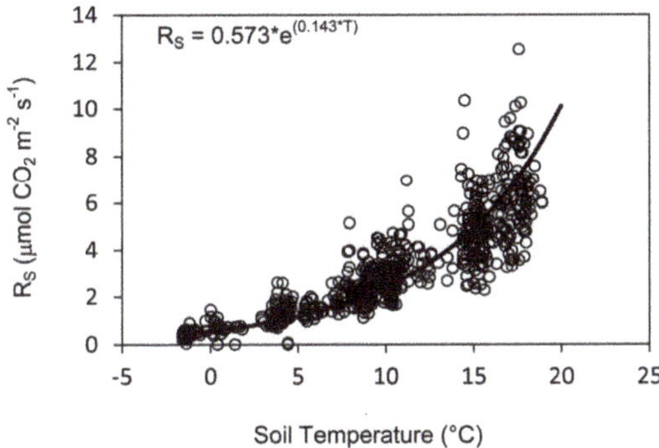

Figure 3. Relationships between soil respiration (R_S) and soil temperature in red pine stands in northern Minnesota, USA.

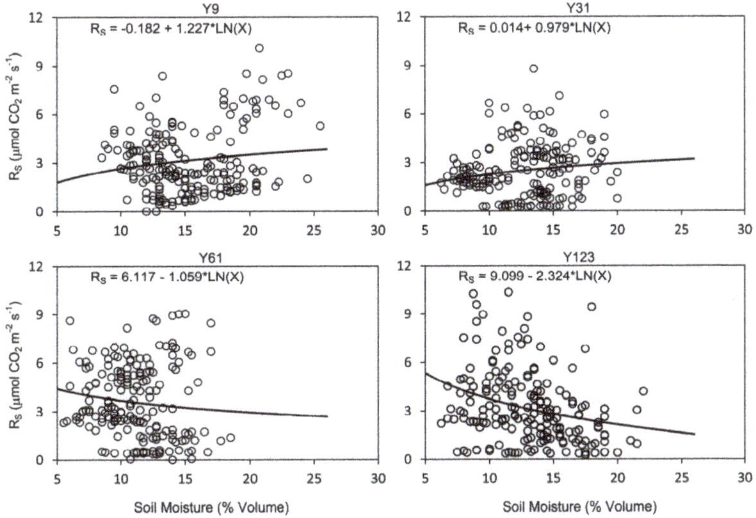

Figure 4. Relationships between soil respiration (R_S) and soil moisture in 9 (Y9), 31 (Y31), 61 (Y61), and 123 (Y123) year-old red pine stands in northern Minnesota, USA.

R_{SN} was not significantly correlated with soil moisture ($p = 0.407$), and age, litter treatment, and the interactions among these three variables did not have significant effects on R_{SN} in any of the models we tested (Figure 5). Further, no R_{SN} models including any combination of soil moisture, age, litter treatment and their interactions performed better than the null model for R_{SN} that included only the intercept and mixed-effects terms accounting for the spatial nesting of collars within plots and the repeated measurements of individual collars and plots over time. R_{SDIF} did have a significant, negative relationship with soil moisture ($p = 0.0137$), but stand age, litter treatment, and their interactions were never significant, and no models including stand age or litter treatment as fixed effects improved on the fit of the general R_{SDIF} to soil moisture model (Figure 5). Further, the correlation between R_{SDIF} and soil moisture was extremely weak ($R^2_C = 0.009$).

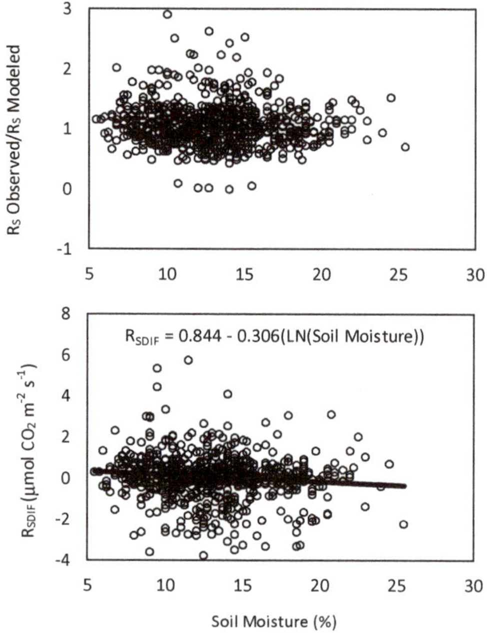

Figure 5. Relationships between two normalized estimates of soil respiration (R_S) and soil moisture in red pine stands in northern Minnesota, USA. R_S Observed (top panel) indicates direct, field-based measurements of soil CO_2 efflux, and R_S modeled (bottom panel) indicates estimates of soil CO_2 efflux derived from our best-fitting regression model relating R_S to temperature. R_{SDIF} was calculated as the differences between R_S Observed and R_S Modeled.

4. Discussion

Our results suggest that stand age may have significant impacts on soil carbon cycling in red pine ecosystems, but stand age had different effects on R_S and R_{FF}. Our young, closed-canopy stand in the early stages of stem exclusion (Y31) showed a distinct seasonal pattern of R_S compared to a young stand regenerating after a clearcut, a mature stand, and an old stand. In contrast, stand age had little impact on R_{FF} or R_{FF}'s contribution to R_S. Our results also suggest that the forest floor influences soil respiration both directly, through the contribution of CO_2 efflux associated with the decomposition of litter material, and indirectly, by altering the physical environment within the upper mineral soil.

4.1. Stand Age Effects on Soil Respiration

Our results do not support the increasing trend of R_S that was predicted based on linkages between productivity and respiration [33]. Although theory predicts a possible decline in R_S from middle-aged to older forests [12], several studies have reported positive relationships between stand age and soil respiration rates spanning various stages of forest development [6,13,16,17]. We found no evidence of either of these patterns, but we did find that soil respiration rates were often lower in Y31 than in the other stands during the middle of the growing season, particularly in July and August when soil temperatures were highest. Y31 often had lower soil temperatures than the other stands, particularly Y9 and Y61. As other studies have indicated [10,19,37], we found that R_S increased exponentially with increasing soil temperature, so low soil temperatures are a likely cause of the low soil respiration rates in Y31. In our study system, Y31 is representative of plantations in the stem exclusion stage of development. Dense canopies that severely limit understory light availability are common characteristics of this phase [38], which could explain the low soil temperatures we observed in Y31. Although we did not directly measure either understory light availability or overstory leaf area index in these stands, we anecdotally observed that Y31 had very limited understory forb and shrub development in comparison with the other sites, which could be indicative of lower light levels on the forest floor, and thus, lower soil temperatures.

Although the R_S-temperature relationship appeared relatively stable across our stands, we did find evidence that the R_S-soil moisture relationship varied among different-aged stands. Some studies have reported that soil respiration's sensitivity to temperature can change with stand age or successional status [13,16,19], but few have reported different age-related trends in the R_S-moisture relationship. Like Irvine and Law [6], we found that soil respiration rates were generally higher in our older stands than in our younger stands during periods of low soil moisture. As soil moisture increased, however, soil respiration rates increased in Y9 and Y31 stands, but declined in Y61 and Y123. While the cause of declining R_S as soil moisture increased in Y61 and Y123 stands is unclear, the effect (i.e., decreasing differences in R_S between young and older stands as soil moisture increases) is consistent with Irvine and Law's [6] proposal that the larger trees in older stands may avoid drought impacts on metabolic processes by accessing water in deeper soil horizons. The possibility of reduced drought impacts on metabolic processes in older stands is also consistent with findings from foliar gas exchange studies that suggest larger trees in older stands show less physiological response to drought than smaller trees in younger stands [39–41]. Thus, the age-related trends we observed for the R_S-moisture relationship are likely driven by more pronounced physiological responses (including respiratory functions associated with growth and maintenance) to drought in younger stands. However, we caution that relationships between R_S and soil moisture in our dataset were extremely weak, so soil moisture availability was not likely a major driver of R_S variability in our system. Soil moisture typically exerts strong controls on R_S in forests only when soils are very dry or very wet [10,42], so it is possible that the 5–15% range of volumetric soil moisture content observed in this study simply did not produce conditions that would limit either decomposer activity or root respiration.

Although we found only limited evidence of age-related variability in the R_S-moisture relationship, seasonal patterns of R_S appeared to closely track soil temperature trends, regardless of stand age. The plateaued pattern of R_S characteristic of most stands in 2009 and the late summer peak in R_S observed in all stands in 2010 paralleled the monthly soil temperature variability for those two years, and the high R_S values observed in Y61 during September 2010 occurred during a period with higher soil temperatures in that stand than in any other stand. In contrast the seasonal dips in soil moisture observed in September of 2009 and June of 2010 did not appear to have a large influence on R_S and the correlation coefficient for our R_S-moisture relationship was an order of magnitude lower than that of our R_S-temperature relationship. Annualized averages of R_S across months of measurement in both years also closely tracked stand-level average soil temperatures, and the generally low soil temperatures in the dense-canopied Y31 appear to be a primary driver of this stand's low R_S, particularly when compared with the young, open-canopied Y9, which had considerably higher average soil temperatures.

Additionally, our two normalized estimates of R_S (i.e., R_{SN} and R_{SDIF}), which expressed observed R_S relative to predictions from our R_S-temperature model, showed little to no relationship with soil moisture. This suggests that some of the apparent soil moisture impacts on R_S across stand of different ages that we observed were likely driven by covarying temperature trends as was observed in other studies e.g., [10,11]. Collectively, these results underscore the importance of soil temperature as a driver of R_S variability e.g., [3,4], and suggest that variations in stand structure that influence soil temperatures (such as the transition from an open-canopy during cohort establishment or stand initiation phase to a dense, closed-canopy during stem exclusion) could potentially exert significant controls on soil carbon cycling.

Although we did not directly test the relationships, variability in soil temperature paralleled differences in temperatures across the two years, while soil moisture generally tracked precipitation levels in the wetter 2010 sampling period, but not as closely in the drier summer of 2009 (Table 4). That both soil temperatures and soil respiration followed the summer plateau pattern of air temperatures in 2009 and the gradual rise in air temperatures through August of 2010, regardless of stand age, speaks to the strong controls that climate exerts on soil processes.

Table 4. Mean monthly temperature (T) and precipitation (P) data for our study area during the months of R_S measurement in 2009 and 2010 [43].

Month	T (°C) 2009	T (°C) 2010	P (mm) 2009	P (mm) 2010
April	4.3	9.1	31.5	28.4
May	10.6	12.6	41.7	59.1
June	16.0	16.9	85.6	91.1
July	17.6	21.1	58.4	120.4
August	17.7	20.8	63.7	129.7
September	17.4	11.7	46.7	141.5
October	3.6	8.8	125.2	66.5

4.2. Forest Floor Contributions to Soil Respiration

Our results indicate that forest floor removal reduced R_S by 4–13% over the course of our April through October sampling (when averaged across R_{FF} estimates from the no litter and removable bag treatment), but do not suggest the anticipated age-related effects on either R_{FF} or the forest floor's contributions to R_S. While several studies have found that respiration from organic horizons represents a larger percentage of R_S than our estimates [18–20,22,44], our results fall within the range reported by others [4,22,24], and support the general consensus across these studies that root respiration and decomposition of organic compounds within the mineral soil represent the largest contributions to R_S.

The reasons for the relatively low forest floor contributions to R_S found in our study are not clear, although substrate quality could be a factor. Like the mineral soil, respiration rates in the forest floor are negatively correlated with the C:N ratio of the substrate [4], and three of our four stands had relatively high forest floor C:N ratios. Forest floor C:N ratios ranged from 39–45 in our Y31, Y61, and Y123 stands. In contrast, forest floor C:N ratios range from 31–32 in young and old-growth Appalachian hardwood forests [4], 34–35 for Douglas-fir (*Pseudotsuga menziesii* Franco) stands in the western United States [24], 29–32 in young to middle-aged Chinese hardwood forests [45], and 21–27 for various components of the forest floor in a tropical montane cloud forest [41]. While the forest floor C:N ratio of our youngest stand was somewhat lower than the other three stands, this stand had a much smaller forest floor layer than the older stands. Forest floor thickness is positively correlated with respiration rates [3,15], so the thinner forest floor may have offset the effects of higher quality litter in Y9, which could explain why R_{FF} estimates were similar across stands despite the difference in forest floor C:N ratios.

The result that R_S in the removable bag treatment was similar to R_S for the control treatments suggests that at least some portion of the difference between control and no litter treatments was due to the influence that forest floor layers exert on the mineral soil environment [31]. Soil temperatures, however, were similar among the three litter treatments. Soil moisture was higher in the no litter treatment, but soil moisture generally has a positive relationship with R_S [5,7], and had little to no relationship with temperature-normalized estimates of R_S in our study, so increased soil moisture is not a likely explanation for lower respiration rates in the no litter treatment. If anything, increased soil moisture in the no litter treatment may have compensated in part for reductions in R_S associated with the absence of a decomposing forest floor. If the difference in soil respiration rates between no litter and control collars was due entirely to the absence of respiration contributions from the forest floor, however, we would expect a similar difference to be apparent in the removable bag treatment. The apparent importance of indirect forest floor impacts on R_S could suggest that R_{FF} and mineral soil respiration estimates from studies that do not account for the forest floor's environmental influences may contain considerable error, at least in systems with relatively thick forest floor layers, such as the red pine forest studied here. The lack of soil temperature differences between the no litter and removable bag treatments, and limited impact of soil moisture on R_S observed at our study sites, however suggest that indirect controls of the soil respiration other than soil microclimate drove the differences in R_{FF} estimates between the no litter and removable bag treatments that we observed.

Negative estimates of R_{FF} during some individual measurement periods were an artifact of our R_{FF} estimation technique, which relied on the difference in soil CO_2 efflux measurements between control collars and adjacent collars where litter was removed. Although efforts were made to place adjacent collars in similar environments and adjacent collars generally had similar soil environments (as measured by soil temperature and moisture), differences in autotrophic contributions to soil CO_2 efflux associated with variable live root biomass beneath collars may have contributed error to our R_{FF} estimates. However estimates of the contribution of R_{FF} to total R_S in a study that used similar methods of estimating R_{FF} while also controlling for root respiration via trenching were broadly similar to our estimates [18], Additionally, litter contributions to total R_S have been shown to be very low during the pre-growth and pre-dormancy period [22], which suggests that even small differences in autotrophic respiration could have contributed to the generally negative estimates of R_{FF} for our earliest and latest measurement periods.

5. Conclusions

Our results have some important implications about C cycling in red pine ecosystems. First, soil respiration responses to changes in temperature appear to be relatively constant across stand ages in red pine. Differences in soil temperature also appear to contribute to low soil respiration rates in dense, young, closed-canopy red pine plantations when compared to open-canopied regenerating stands and older stands that have transitioned out of the stem exclusion phase of development. This suggests that changes in rotation lengths for red pine management could impact soil C dynamics by modifying the proportional representation of age classes across the landscape and consequently, shifting the relative abundance of stand structures with relatively low soil respiration rates (e.g., our young, closed canopy plantation) vs. those with higher, or more seasonally-variable soil respiration rates (e.g., our young, open-canopied stand and our more mature stands).

Acknowledgments: We would like to thank Doug Kastendick, Nate Aspelin, Eric Doty, and Ryan Kolka for their assistance in locating field sites, installing respiration collars and litter bags, collecting soil respiration data, and processing litter samples in the laboratory. Funding for this project was provided by the United States Department of Agriculture Forest Service Northern Research Station and the Agenda 2020 Program. Any use of trade, product, or firm names is for descriptive purposes only and does not imply endorsement by the United States Government.

Author Contributions: R.K., J.B., B.P. and M.J. conceived and designed the experiments; M.P. performed the experiments and analyzed the data; M.P. wrote the paper with significant feedback from R.K., J.B., B.P. and M.J.

Conflicts of Interest: The authors declare no conflict of interest.

References

1. Law, B.E.; Ryan, M.G.; Anthoni, P.M. Seasonal and annual respiration of a ponderosa pine ecosystem. *Glob. Chang. Biol.* **1999**, *5*, 169–182. [CrossRef]

2.	Janssens, I.A.; Lankreijer, H.; Matteucci, G.; Kowalski, A.S.; Buchmann, N.; Epron, D.; Pilegaard, K.; Kutsch, W.; Longdoz, B.; Gruènwald, T.; et al. Productivity overshadows temperature in determining soil and ecosystem respiration across European forests. *Glob. Chang. Biol.* **2001**, *7*, 269–278. [CrossRef]

3.	Saiz, G.; Green, C.; Butterback-Bahl, K.; Kiese, R.; Avitabile, V.; Farrell, E.P. Seasonal and spatial variability of soil respiration in four Sitka spruce stands. *Plant Soil* **2006**, *287*, 161–176. [CrossRef]

4.	Vose, J.M.; Bolstad, P.V. Biotic and abiotic factors regulating forest floor CO_2 flux across a range of forest age classes in the southern Appalachians. *Pedobiologia* **2007**, *50*, 577–587. [CrossRef]

5.	Orchard, V.A.; Cook, F.J. Relationship between soil respiration and soil moisture. *Soil Biol. Biochem.* **1983**, *15*, 447–453. [CrossRef]

6.	Irvine, J.; Law, B.E. Contrasting soil respiration in young and old-growth ponderosa pine forests. *Glob. Chang. Biol.* **2002**, *8*, 1183–1194. [CrossRef]

7.	Cook, F.J.; Orchard, V.A. Relationships between soil respiration and soil moisture. *Soil Biol. Biochem.* **2008**, *40*, 1013–1018. [CrossRef]

8.	Raich, J.W.; Nadelhoffer, K.J. Belowground carbon allocation in forest ecosystems: Global trends. *Ecology* **1989**, *70*, 1346–1354. [CrossRef]

9.	Søe, A.R.B.; Buchmann, N. Spatial and temporal variations in soil respiration in relation to stand structure and soil parameters in an unmanaged beech forest. *Tree Physiol.* **2005**, *25*, 1427–1436. [CrossRef] [PubMed]

10.	Davidson, E.E.; Belk, E.; Boone, R.D. Soil water content and temperature as independent or confounded factors controlling soil respiration in a temperate mixed hardwood forest. *Glob. Chang. Biol.* **1998**, *4*, 217–227. [CrossRef]

11.	Wang, B.; Zha, T.S.; Jia, X.; Wu, B.; Zhang, Y.Q.; Qin, S.G. Soil moisture modified the response of soil respiration to temperature in a desert shrub ecosystem. *Biogeosciences* **2014**, *11*, 259–268. [CrossRef]

12.	Pregitzer, K.S.; Euskirchen, E.S. Carbon cycling and storage in world forests: Biome patterns related to forest age. *Glob. Chang. Biol.* **2004**, *10*, 1–26. [CrossRef]

13.	Tang, J.; Bolstad, P.V.; Martin, J.G. Soil carbon fluxes and stocks in a Great Lakes forest chronosequence. *Glob. Chang. Biol.* **2009**, *15*, 145–155. [CrossRef]

14.	Wang, B.; Jiang, Y.; Wei, X.; Zhao, G.; Guo, H.; Bai, X. Effects of forest type, stand age, and altitude on soil respiration in subtropical China. *Scand. J. For. Res.* **2011**, *26*, 40–47. [CrossRef]

15.	Saiz, G.; Byrne, K.A.; Butterbach-Bahl, K.; Kiese, R.; Blujdea, V.; Farrell, E.P. Stand age-related effects on soil respiration in a first rotation Sitka spruce chronosequence in central Ireland. *Glob. Chang. Biol.* **2006**, *12*, 1007–1020. [CrossRef]

16.	Martin, D.; Beringer, J.; Hutley, L.B.; McHugh, I. Carbon cycling in a mountain ash forest: Analysis of below ground respiration. *Agric. For. Meteorol.* **2007**, *147*, 58–70. [CrossRef]

17.	Gough, C.M.; Seiler, J.R.; Wiseman, P.E.; Maier, C.A. Soil CO_2 efflux in loblolly pine (*Pinus taeda* L.) plantations on the Virginia Piedmont and South Carolina Coastal Plain over a rotation-length chronosequence. *Biogeochemistry* **2005**, *73*, 127–147. [CrossRef]

18.	Bowden, R.D.; Nadelhoffer, K.J.; Boone, R.D.; Melillo, J.M.; Garrison, J.B. Contributions of aboveground litter, belowground litter, and root respiration to total soil respiration in a temperate mixed hardwood forest. *Can. J. For. Res.* **1993**, *23*, 1402–1407. [CrossRef]

19.	Buchmann, N. Biotic and abiotic factors controlling soil respiration rates in *Picea abies* stands. *Soil Biol. Biochem.* **2000**, *32*, 1625–1635. [CrossRef]

20.	Borken, W.; Beese, F. Soil respiration in pure and mixed stands of European beech and Norway spruce following removal of organic horizons. *Can. J. For. Res.* **2005**, *35*, 2756–2764. [CrossRef]

21.	Ngao, J.; Epron, D.; Brechet, C.; Granier, A. Estimating the contribution of leaf litter decomposition to soil CO_2 efflux in a beech forest using ^{13}C-depleted litter. *Glob. Chang. Biol.* **2005**, *11*, 1168–1776. [CrossRef]

22.	DeForest, J.L.; Chen, J.; McNulty, S.G. Leaf litter is an important mediator of soil respiration in an oak-dominated forest. *Int. J. Biometeorol.* **2009**, *53*, 127–134. [CrossRef] [PubMed]

23.	Martin, J.G.; Bolstad, P.V. Variation of soil respiration at three spatial scales: Components within measurements, intra-site variation and patterns on the landscape. *Soil Biol. Biochem.* **2009**, *41*, 530–543. [CrossRef]

24.	Giesen, T.W.; Perakis, S.S.; Cromack, K., Jr. Four centuries of soil carbon and nitrogen change after stand-replacing fire in a forest landscape in the western Cascade Range of Oregon. *Can. J. For. Res.* **2008**, *38*, 2455–2464. [CrossRef]

25. Bradford, J.B.; Kastendick, D.N. Age-related patterns of forest complexity and carbon storage in pine and aspen-birch ecosystems of northern Minnesota, USA. *Can. J. For. Res.* **2010**, *40*, 401–409. [CrossRef]

26. Borken, W.; Davidson, E.A.; Savage, K.; Gaudinxki, J.; Trumbore, S.E. Drying and wetting effects on CO_2 release from organic horizons. *Soil Sci. Soc. Am. J.* **2003**, *67*, 1888–1896. [CrossRef]

27. Sulzman, E.W.; Brant, J.B.; Bowden, R.D.; Lajtha, K. Contribution of aboveground litter, belowground litter, and rhizosphere respiration to total soil CO_2 efflux in an old growth coniferous forest. *Biogeochemistry* **2005**, *73*, 231–256. [CrossRef]

28. Mendham, D.S.; O'Connell, A.M.; Grove, T.S.; Rance, S.J. Residue management effects on soil carbon and nutrient contents and growth of second rotation eucalypts. *Forest Ecol. Manag.* **2003**, *181*, 357–372. [CrossRef]

29. Walmsley, J.D.; Godbold, D.L. Stump harvesting for bioenergy—A review of the environmental impacts. *Forestry* **2010**, *83*, 17–38. [CrossRef]

30. Peckham, S.D.; Gower, S.T. Simulated long-term effects of harvest and biomass residual removal on soil carbon and nitrogen content and productivity for two Upper Great Lakes forest ecosystems. *Glob. Chang. Biol. Bioenergy* **2011**, *3*, 135–147. [CrossRef]

31. Sayer, E.J. Using experimental manipulation to assess the roles of leaf litter in the functioning of forest ecosystems. *Biol. Rev.* **2006**, *81*, 1–31. [CrossRef] [PubMed]

32. D'Amato, A.W.; Palik, B.J.; Kern, C.C. Growth, yield, and structure of extended rotation *Pinus resinosa* stands in Minnesota, USA. *Can. J. For. Res.* **2010**, *40*, 1000–1010. [CrossRef]

33. Högberg, P.; Nordgren, A.; Buchmann, N.; Taylor, A.F.S.; Ekblad, A.; Högberg, M.N.; Nyberg, G.; Ottosson-Löfvenius, M.; Read, D.J. Large-scale forest girdling shows that current photosynthesis drives soil respiration. *Nature* **2001**, *411*, 789–792. [CrossRef] [PubMed]

34. Nyberg, P.R. *Soil survey of Itasca County, Minnesota*; USDA Soil Conservation Service Government Printing Office: Washington, DC, USA, 2008.

35. Powers, M.D.; Kolka, R.K.; Bradford, J.B.; Palik, B.J.; Fraver, S.; Jurgensen, M.F. Carbon storage across a chronosequence of thinned and unmanaged red pine stands. *Ecol. Appl.* **2012**, *22*, 1297–1307. [CrossRef] [PubMed]

36. Huang, S.; Meng, S.X.; Yang, Y. Assessing the goodness of fit of forest models estimated by nonlinear mixed-model methods. *Can. J. For. Res.* **2009**, *39*, 2418–2436. [CrossRef]

37. Boone, R.D.; Nadelhoffer, K.J.; Canary, J.D.; Kaye, J.P. Roots exert a strong influence on the temperature sensitivity of soil respiration. *Nature* **1998**, *396*, 570–572. [CrossRef]

38. Oliver, C.D.; Larson, B. *Forest Stand Dynamics*; Wiley: New York, NY, USA, 1996.

39. McDowell, N.G.; Licata, J.; Bond, B.J. Environmental sensitivity of gas exchange in different-sized trees. *Oecologia* **2005**, *145*, 9–20. [CrossRef] [PubMed]

40. Ryan, M.G.; Phillips, N.; Bond, B.J. The hydraulic limitation hypothesis revisited. *Plant Cell Environ.* **2006**, *29*, 367–381. [CrossRef] [PubMed]

41. Wharton, S.; Schroeder, M.; Bible, K.; Falk, M.; Paw U, K.T. Stand-level gas-exchange responses to seasonal drought in very young versus old Douglas-fir forests of the Pacific Northwest, USA. *Tree Physiol.* **2009**, *29*, 959–974. [CrossRef] [PubMed]

42. Borken, W.; Xu, Y.; Davidson, E.A.; Beese, F. Site and temporal variation of soil respiration in European beech, Norway spruce, and Scots pine forests. *Glob. Chang. Biol.* **2002**, *8*, 1205–1216. [CrossRef]

43. NOAA National Climatic Data Centers Climate Data Online. Available online: https://www.ncdc.noaa.gov/cdo-web/ (accessed on 5 December 2017).

44. Zimmermann, M.; Meir, P.; Bird, M.; Malhi, Y.; Cahuana, A. Litter contribution to diurnal and annual soil respiration in a tropical montane cloud forest. *Soil Biol. Biochem.* **2009**, *41*, 1338–1340. [CrossRef]

45. Yang, X.; Chen, J. Plant litter quality influences the contribution of soil fauna to litter decomposition in humid tropical forests, southwestern China. *Soil Biol. Biochem.* **2009**, *41*, 910–918. [CrossRef]

 *forests*

Article

Carbon Sequestration in Protected Areas: A Case Study of an *Abies religiosa* (H.B.K.) Schlecht. et Cham Forest

Pablo I. Fragoso-López [1], Rodrigo Rodríguez-Laguna [2], Elena M. Otazo-Sánchez [1], César A. González-Ramírez [1], José René Valdéz-Lazalde [3], Hermann J. Cortés-Blobaum [1] and Ramón Razo-Zárate [2,*]

[1] Área Académica de Química, Instituto de Ciencias Básicas e Ingeniería, Universidad Autónoma del Estado de Hidalgo, Kilómetro 4.5 carretera Pachuca—Tulancingo, Mineral de la Reforma, Hidalgo 42090, Mexico; fr342891@uaeh.edu.mx (P.I.F.-L.); profe_4339@uaeh.edu.mx (E.M.O.-S.); cramirez@uaeh.edu.mx (C.A.G.-R.); co342890@uaeh.edu.mx (H.J.C.-B.)

[2] Área Académica de Ciencias Agrícolas y Forestales, Instituto de Ciencias Agropecuarias, Universidad Autónoma del Estado de Hidalgo, Av. Universidad km. 1, Ex. Hda de Aquetzalapa, Tulancingo, Hidalgo 43600, Mexico; rlaguna@uaeh.edu.mx

[3] Colegio de Postgraduados, Km 36.5 Carr. México-Texcoco, Montecillo, Texcoco 56230, Mexico; valdez@colpos.mx

* Correspondence: ramon_razo@uaeh.edu.mx; Tel.: +52-771-71-72000 (ext. 2431)

Received: 23 August 2017; Accepted: 28 October 2017; Published: 12 November 2017

Abstract: The effects of global climate change have highlighted forest ecosystems as a key element in reducing the amount of atmospheric carbon through photosynthesis. The objective of this study was to estimate the amount of carbon content and its percentage capture in a protected *Abies religiosa* forest in which the study area was zoned with satellite image analysis. Dendrometric and epidometric variables were used to determine the volume and increase of aerial biomass, and stored carbon and its capture rate using equations. The results indicate that this forest contains an average of 105.72 MgC ha^{-1}, with an estimated sequestration rate of 1.03 MgC ha^{-1} yr^{-1}. The results show that carbon capture increasing depends on the increase in volume. Therefore, in order to achieve the maximum yield in a forest, it is necessary to implement sustainable forest management that favors the sustained use of soil productivity.

Keywords: climate change; protected forest; carbon sequestration; *Abies religiosa*

---

1. Introduction

The negative impacts of anthropogenic emissions of greenhouse gases such as irreversible damage to ecosystems, increased pressure on water resources, alterations in food production, and damage to human health, among others, have been reported in different studies [1–10]. The need to stabilize the carbon content of the atmosphere has been manifested in a series of international and local agreements and policies, such as the Kyoto Protocol and the Treaty of Paris. The purpose of these agreements and policies is to reduce emissions of greenhouse gases (GHG), with mechanisms to optimize carbon sinks.

Currently, forests store about 800 gigatons of carbon (GtC) [11] and it is estimated that by 2050 they could sequester up to an additional 87 GtC [12,13]. It was estimated that in the period between 2000 and 2007, the carbon sequestration rate of the world's forests averaged 4.1 GtCyr^{-1} [14], corresponding to approximately 30% of fossil fuel emissions in 2010 [15].

Globally, protected forests have been proposed as a potentially cost-effective strategy to counter deforestation and degradation [16–18], favoring carbon permanence in the forest. Countries with the greatest threats from their forests due to degradation and devastation have increased their percentage of protected areas (Figure 1) in an attempt to conserve the environmental services of their forests [19,20]. Out of 3984 million hectares of forests in the world, 13.25% have a protected area status [21], and this percentage is mainly because of many of these protected sites partially fulfilling their conservation objectives [22], primarily derived from the budgetary constraints in which most of these areas operate. Financial resources managers for protected areas are increasingly emphasizing cost-effective aspects such as ecosystem services, including carbon sequestration [23].

Figure 1. Percentage of protected area per country in 2015 in relation to its total forest area. Own elaboration with information of [24].

Currently, decision-makers can use a large number of methods to assess protected area management and prioritize investments, and there are over 70 methods that have been developed to provide standards for obtaining indicators, analyses, and interpretations [25,26]; however, none of these provide an estimation of carbon sequestration rates.

The present study aimed to estimate the carbon content in the above-ground biomass (baseline) of a forest of *Abies religiosa* (H.B.K.) Schlecht. et Cham., which is part of El Chico National Park, Hidalgo, Mexico. In addition, it attempted to ascertain the carbon capture rate from the annual volumetric increase of the species in the area. This information is necessary for designing the strategies that allow the environmental objectives that contribute to the reduction of the negative effects in climate change to be fulfilled.

2. Materials and Methods

2.1. Study Area

The study was carried out in the federal zone of the protected natural area called "El Chico National Park" located in the western area of the mountain range of Pachuca, in the state of Hidalgo, Mexico. Geographically, it is located between the extreme coordinates 20°11'57" and 20°12'02" north latitude and 98°43'08"and 98°43'06" west longitude.

The area is owned by the Mexican nation and comprises 1,833,000 hectares [27]. The climate is C (m) (w) b (i') gw, which is described as: a temperate-sub-humid climate with a fresh and long summer; the average annual temperature varies between 12 and 18 °C. The rainfall regime is in summer, and the percentage of winter rain in relation to the annual total is less than 5% [28].

The soil type is constituted by associations that group the following soil units: Humic cambisol-Androsol ocrico-Litosol, which has a volcanic origin association typical of mountainous zones, and humic Andosol-Humic cambisol, corresponding to forest soils associated with *Abies* and *Quercus* forests [29].

The predominant type of vegetation is *Abies religiosa* forest, which covers 67% of the park surface. The shrub stratum is dominated by the species *Archibaccharis hieracioides* Blake, *Baccharis conferta* H.B.K., *Eupatorium hidalgense* Rob., *Fuchsia thymifolia* H.B.K., *Ribes affine* H.B.K., *Salvia elegans* Vahl., *Senecio angulifolius* D.C., and *Stevia monardifolia* H.B.K. [30].

2.2. Description of the Species

The *Abies* forest is considered a component of the Leopold boreal forest [31] because of its similarity in terms flora, fauna, physiognomy, and ecological conditions to large forest masses covering northern parts of North America and Eurasia, and is also known as Taiga. This type of forest usually develops in an interval of altitude between 2400 and 3600 m, and its humidity requirement is high, registering precipitations superior to 1000 mm annually. They are dense forests, of heights that can reach up to 50 m, with canopies of a triangular contour; the high density conditions reduce the amount of light reaching the interior, which limits the development of shrubs and herbaceous species [32].

2.3. Forest Zoning

This process was performed to identify the areas where the species of interest is dominant. The stands were also identified where the species is mixed with other genera; however, these were excluded from the study. This concept is also known as stratification and consists of the division of the forest area into portions or spatial units called stands, with sections that possess similar physical and biological characteristics [33,34].

The identification of the stands was made through a RapidEye-4 satellite (provided by the Space Agency of the Mexican Government) image analysis dated 25 February 2015, with a spatial resolution of 5 m [35]. Initially, the image was corrected atmospherically and radiometrically [36]. Subsequently, it was estimated at the pixel level "Red Edge Normalized Difference Vegetation Index" ($RedEdge_{NDVI}$) [37–41].

$$RedEdge_{NDVI} = \frac{R710 - R705}{R710 + R705} \tag{1}$$

where R710 is the band 4 RedEdge or near red and R705 corresponds to the band 3 RedEdge. The NDVI has been used in the elaboration of the process of forest logging, and it is mentioned that it has shown acceptable results in the identification of vegetal associations.

The RedEdge NDVI results were analyzed along with altitude, latitude, and exposure values to determine the final logging through an overlay positioning process.

The obtained image analysis was validated with the records obtained through field trips. The existing vegetation types and the boundaries between them were determined by georeferenced points.

2.4. Field Information

Sampling Design and Characteristics of Sites

A random sampling design was used to define the location of 33 circular sampling sites of 1000 m^2, in which a total of 682 trees were measured. The shape and size of the sites were suitable for the purpose pursued since they have shown good results for the calculation of volumetric stocks or biomass content [42,43].

At each sampling site, the diameter information at the breast height (dap) of all trees with a diameter equal to or greater than 7.5 cm and the total height of each tree [44] was recorded. For the determination of the number of annual growth rings in 2.5 cm, known as passage time, an average tree per diameter category was selected at each sampling site by means of a 250-mm Haglöf® Presser (Haglöf, Langsele, Sweden) drill and the results were calculated as the annual volumetric increase per hectare [45], which was subsequently used for estimating carbon sequestration.

2.5. Information Processing

2.5.1. Calculation of Volumetric Stocks

With the records of each stand, we proceeded to estimate the value of the variables basal area and timber volume [36,46]. The basal area for each tree was obtained by the following equation:

$$B_A = \frac{\pi}{4} * D_{bh}^2 \tag{2}$$

where B_A is the basal area (m^2) and D_{bh} corresponds to the diameter at breast height (m), which were grouped in diametric categories of 5 cm.

The volume per hectare per stand was determined through a procedure known in the forest as the average hectare [47] by employing the following equation:

$$\overline{Vs} = \sum_{i=1}^{DC} \left(B_{Ai} * \overline{H} * M_{Ci} * E_{Fi} * N_{Ti} \right) \tag{3}$$

where $\overline{Vs}$ corresponds to the total volume (trunk + branches + leaves + roots) of the woodland per hectare (m^3); DC is the number of diametric categories in the stand; i is the diameter category; B_A is the basal area of the average tree by diameter category (m^2); $\overline{H}$ corresponds to the average height of the trees in the stands for each diameter category (meters); M_C is the morphic coefficient for gender Abies; E_F is the expansion factor of stem to total volume (branches, leaves, and roots) [48]; and N_{Ti} is the number of trees per hectare per diameter category.

2.5.2. Volumetric Increment

The calculation of the increment of the forest mass for each of the stands was obtained by means of the "Klepac Fast" method, which relates the number of trees per diameter category per hectare, the volume of the tree type for each diameter category, and the step time for each case [45,49]. This method considers a series of calculations between these three variables to obtain the percentage of the increase by the following equation:

$$P = \frac{1000}{Dbh} * \frac{1}{T} \tag{4}$$

where P corresponds to the percentage of increase, Dbh is the diameter at breast height, and T is the step time. It is worth mentioning that this parameter was used to obtain the carbon sequestration rate.

2.5.3. Aerial Biomass Content

This information was obtained by the equation developed by Avendaño et al. [50], for the calculation of biomass for *Abies religiosa* based on the diameter at breast height.

$$B = 0.0713\, D_{bh}^{2.5104} \tag{5}$$

where B is the total biomass of the tree (Mg) and Dbh is the diameter at breast height (m).

2.5.4. Carbon Content

The estimated carbon content for each tree was calculated using the equation developed for this species, which is based on the diameter at breast height [50].

$$A_{CC} = 0.0332\, D_{bh}^{2.5104} \tag{6}$$

where A_{CC} is the carbon content for *Abies religiosa* (Mg) and D_{bh} is the diameter at breast height (m).

The carbon content for each diameter category was obtained by the following equation:

$$Dc_{cc} = A_{cc} * N_T \tag{7}$$

where Dc_{cc} is the carbon content by diameter category (Mg), A_{cc} is the carbon content per tree (Mg), and N_T is the number of trees of the corresponding diametric category.

The carbon content per hectare was obtained by adding the carbon content from all the diametric categories.

2.5.5. Carbon Sequestration Rate

After calculating the carbon content per hectare and after having determined the volumetric increments, the carbon capture rate for each stand was determined, considering that the increment parameter refers to the volume increase per unit time. Once the amount of biomass that this forest can generate in a certain time period was obtained, the carbon stored was calculated using the following equation:

$$\overline{C}_{SR} = (\overline{C}_C / \overline{V}s) * C_{AI} \tag{8}$$

where $\overline{C}_{SR}$ is the rate of carbon sequestration per hectare per year (Mg), $\overline{C}_C$ is the carbon content (MgC ha^{-1}), $\overline{V}s$ corresponds to the volumetric stocks (m^3 ha^{-1}), and C_{AI} is the current annual increase (m^3).

3. Results

3.1. Forest Zoning and Field Information

Of the total area studied, eight stands were identified, with *Abies religiosa* covering an area of 1229.65 hectares. The rest of the area (603.35 hectares) corresponds to rock formations and other types of vegetation such as *Quercus*, *Juniperus*, and grassland. The study was addressed to *Abies religiosa* since it is the conservation species of interest in this protected area (Figure 2).

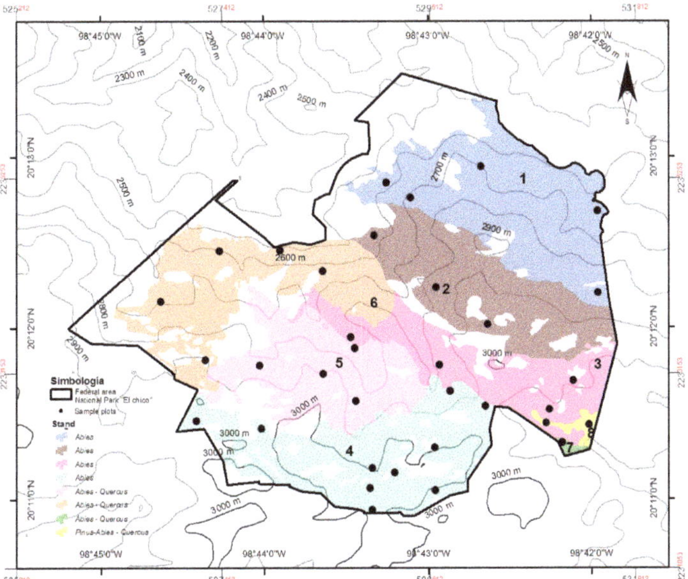

Figure 2. Forest zoning obtained through the analysis of satellite images in combination with altitude information, orientation, and slope.

The Table 1 show the values of the NDVI together with the land orientation, as well as the slope range that exists in each one of the stands, for which the criteria for the realization of the forest zoning were the values of NDVI, orientation, and species composition.

Table 1. Criteria and values used for forest zoning.

Stand	NDVI			Orientation	Slope	Species Composition
	Mean	Min	Max			
1	0.3495	0.1764	0.4350	North	10–25°	*Abies*
2	0.3709	0.2043	0.4681	North	10–25°	*Abies*
3	0.3678	0.3016	0.4806	Southeast	10–25°	*Abies*
4	0.3900	0.2004	0.4613	Northwest	10–25°	*Abies*
5	0.3849	0.2773	0.4510	Northwest	10–25°	*Abies-Quercus*
6	0.3603	0.2751	0.4360	North	10–25°	*Abies-Quercus*
7	0.3849	0.3228	0.4228	East	10–25°	*Abies-Quercus*
8	0.3441	0.2841	0.4064	Northwest	4–9°	*Abies-Pinus-Quercus*
Other *	0.2753	0.0228	0.3990	-	-	-

* Refers to other uses or plant formations, "-" refers to not obtained.

3.2. Volumetric Stocks

The total volume is 757,681.69 m^3, and the average volumetric stock per hectare is 616.18 m^3. Table 2 details the *Abies religiosa* volumes for each stand.

Table 2. Volume in m^3 per stand.

Stand	Surface (Hectares)	Number of Trees ha^{-1}	Basal Area m^2 ha^{-1}	Volumetric Stocks m^3 ha^{-1}	Volumetric Stocks m^3 Stndl^{-1}
1	263.43	220	47.37	771.57	203,252.50
2	194.20	210	35.18	716.46	139,139.25
3	153.63	210	32.63	672.41	103,299.63
4	223.27	282	40.44	789.37	176,238.35
5	178.26	140	22.02	462.22	82,392.82
6	203.92	70	11.86	239.33	48,805.67
7	3.29	270	36.84	727.16	2524.82
8	9.66	140	12.37	223.86	1770.21

Stands 1, 2, 3, 4, and 7 have a density of more than 200 trees per hectare, a basal area greater than 30 m^2, and a volume that exceeds 670 m^3 ha^{-1}. Stands 5, 6, and 8 have densities of less than 150 trees per hectare and volumes below 470 m^3 ha^{-1}, which is due to the presence of disturbances such as forest fires and pests that have affected the density per unit area.

3.3. Volumetric Increment

Table 3 presents the results of the calculation of volumetric increase, for which we considered the records of 165 trees. It can be observed that stands 4, 5, and 6 present smaller times of passage in relation to the rest of the stands, probably due to the age and density of the forest mass. The highest productivity is found in stands 1, 4, and 5, where the area of these stands also influences the result.

Table 3. Result of calculating volumetric increments.

Stand	Sampling Plot	Step Time Average	Drilled Trees	Current Annual Increase	
				m^3	%
1	4	19	20	6.616	3.007
2	3	22	15	4.440	2.114
3	4	24	20	4.870	2.319
4	9	17	45	7.077	2.508
5	5	16	25	5.261	3.758
6	4	15	20	2.605	3.722
7	2	23	10	4.662	1.554
8	2	26	10	1.164	0.831

3.4. Carbon Content

The estimate of the total carbon content is 130,004.02 Mg, with an average per hectare of 105.72 MgC. The results for each stand are shown in Table 4.

Table 4. Carbon content per hectare and per stand in megagrams.

Stand	Surface ha^{-1}	Content of C ha^{-1} (MgC)	Content of C Stand^{-1} (MgC)
1	263.43	164.25	43,268.71
2	194.20	110.92	21,540.97
3	153.63	100.44	15,430.38
4	223.27	129.55	28,925.22
5	178.26	70.95	12,646.44
6	203.92	36.25	7392.48
7	3.29	120.68	396.63
8	9.66	41.72	403.18

3.5. Carbon Sequestration

The result for the carbon capture rate in the *Abies religiosa* aerial biomass is 1267.66 MgC yr^{-1}, considering the studied area, with an average of 1.03 MgC yr^{-1} per hectare.

Stands 1 and 4 have higher volumetric increments. Consequently, the sequestered carbon is higher in relation to the rest of the stands (Figure 3), and this can be attributed to the north orientation of the surface where the insolation is less and more moisture is available. Stand 6 is located in a transition zone with the presence of other tree genres that were not counted but that compete with the *Abies*, decreasing its density, due to the fact that the volume and increase of biomass is reduced. Stand 8 presents a situation similar to the previous stand, only in this case the mixture of genera is due to anthropic activities (reforestation).

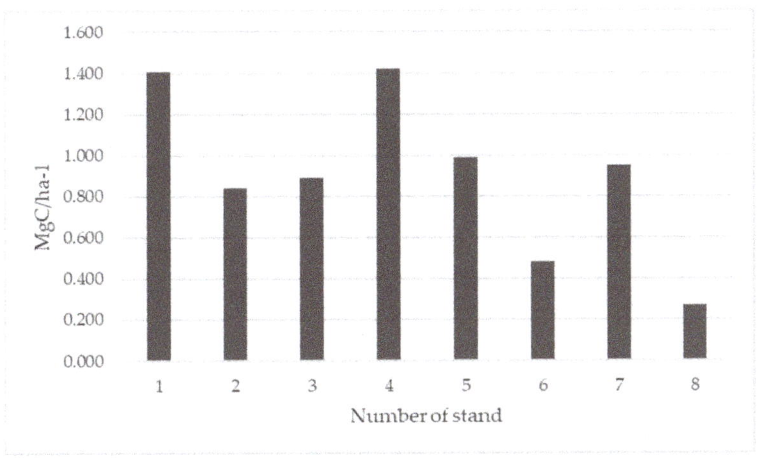

Figure 3. Annual carbon sequestration per hectare in megagrams.

Carbon sequestration in stand 1 is the one with the highest catch mainly due to the surface area and annual increment, followed by stand 4. The strata with the least carbon sequestration are the ones with the lowest surface area and the lowest density of trees (Figure 4).

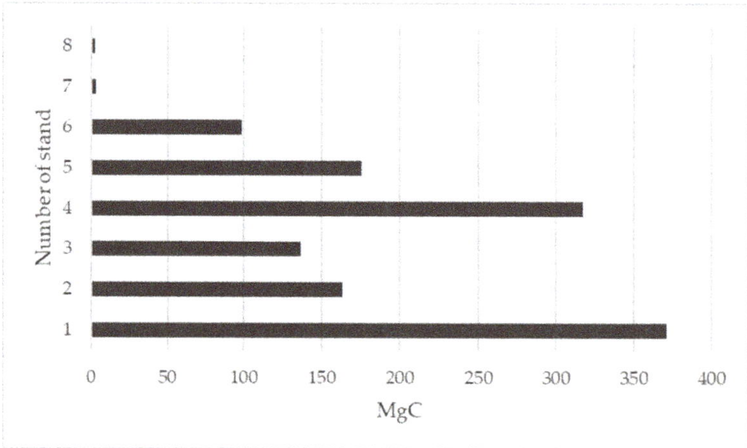

Figure 4. Annual carbon capture above-ground per stand.

4. Discussion

4.1. Carbon Content

The area of study corresponds to a protected area decreed in 1898, in which the conservation policies that prevent the modification of the forest structure have caused the longevity of this mass, placing it in the last stage of development (old fustal), and this is derived from the age and diameters present. Diameters greater than 1.10 m were recorded, while the smallest diameter registered was 7.5 cm. On average, the number of trees per hectare is 193, with a volume of 616.18 m^3. Table 5 presents comparative information of estimated carbon in protected areas with the presence of *Abies religiosa*.

Table 5. Comparison of studies in protected forests where they have calculated the carbon content in aerial biomass in *Abies religiosa*.

Author	Place of Study	Status of the Area	Species	Carbonor Estimated Mg ha^{-1}	Observations
Present study	Hidalgo, Mexico	Protected	*Abies religiosa*	105.72	Long-lived forest mass and scarce natural regeneration.
[51]	Mexico City, Mexico	Protected	*Abies religiosa*	117.00	The author makes reference to 3 associations *Abies religiosa* with shrub and/or herbaceous species. It is not mentioned the method to determine the carbon content.
[52]	Mexico City, Mexico	Protected	*Abies religiosa*	136.41	A conserved forest was studied.
[53]	Hidalgo, Mexico	Protected	*Abies religiosa*	138.62	The author details 3 carbon scenarios for this type of forest, in which include disturbed areas, the area studied was 212.95 hectares.
[54]	Veracruz, Mexico	Protected	Various conifers	146.30	It is a protected forest, the authors detail scenarios where the species of *Abies religiosa* is combined with other conifers.
[55]	Mexico State, Mexico	Protected	*Abies religiosa*	163.62	A similar methodology was used to calculate the carbon content.

There are several hypotheses about the carbon content in forest ecosystems. This storage is dependent on the amount of existing biomass, which depends on the age, diameter, and height of the trees. When woodland density is affected or altered for some cause such as: pests, diseases, forest fires, clandestination, or harvesting, the carbon content per unit area is also altered [56]. The silviculously managed forests have young stands as a result of the application of silvicultural treatments [57], where the stands with a higher density and basal area are those containing more volume of above-ground biomass, and consequently, more carbon content.

The records consulted about carbon contents in protected temperate forest are ≥ 115 MgC ha^{-1} [51–55], whereas this study estimated more than 105.72 MgC ha^{-1}. These differences can be attributed to several factors such as the quality of the site studied, and the density and age of the trees, among others. Unlike the preserved forests, disturbed forests contain less carbon, due to the affected forest mass. Aguirre et al. [57] mentions that for a managed forest of *Pinus patula* with an orientation that is contemporary, the content is 63.98 MgC ha^{-1}.

In the case of protected forests or silvically managed forests, the ecosystem service of carbon storage is fulfilled, whereas in the case of harvested forests, the carbon content is lower and mainly depends on applied silvicultural treatments. For the production of coetaneous masses, the content is variable and dependent on the age, diameter, height, and density of the trees in a given stand. The present study considers there to be 48% more carbon content in El Chico than the data provided by Aguirre et al. [57]. Masera et al. [58] mentions that managed forests with a temperate climate contain 118 MgC ha^{-1} or 10% more than the estimated data in this study. The differences between Aguirre and Masera information can be mainly attributed to the applied silvicultural system (intensive or conservative).

4.2. Carbon Sequestration

Forest ecosystems sequester carbon and are considered as an option for the mitigation of the effects of an increasing atmospheric CO_2 load [59–62]. The potential for carbon of any forest species depends on the maximum amount of biomass it can produce per unit of time. In species of accelerated growth, this parameter is relatively fast reached, whereas in species with a slow growth, the period of time required to reach the maximum biomass content is longer, and consequently, carbon sequestration is higher [63].

The volumetric increases represent the parameter which is able to determine the rate of carbon sequestration, and for the case presented in this study, the *Abies religiosa* forest captures 1.03 MgC ha^{-1} yr^{-1}, which is equivalent to 3.78 MgCO$_2$ ha^{-1} [64], with an annual average increase of 2.92%. Because it is located within a protected area, the forest mass has not been altered, which has caused the increment curve to decline. As the age of the mass increases, the increments decrease [45], and as a consequence, the potential for carbon capture is also affected.

Compared with protected forests, sustainably harvested forest areas capture more carbon [65], which is mainly due to the management of the age factor within the masses. Similarly, the use of harvested biomass for the production of long-lasting products retains carbon for long periods of time [66,67]. There are records of carbon capture in managed forests that exceed the results obtained in this study. Liu et al. [68] estimated net biomass productivity for forests in the Appalachian region where the forest harvest exists, and reported data ranging from 1.8 to 6.2 MgC ha^{-1}. Zhang et al. [69] estimated a value of 2.4 MgC ha^{-1} captured for a Massonian *Pinus* forest in China. For the specific case of *Abies religiosa* with the information of Manzanilla et al. [70], we can estimate 3.1 MgC ha^{-1}. The differences between these investigations and the estimated rate in this study can be attributed to factors such as: forest management, location of the studied area, and species, among others. Navarro et al. [65] mentioned that these results cannot be cofrontable since they correspond to different ecosystems, each with their own particularities.

Unlike carbon content, the rate of sequestration is influenced by forest management techniques; in protected forests, the amount of stored carbon is higher, but the catch rate is reduced. With sustainable forest management practices, this difference can be balanced.

Forest management should be included in protected forests, and it is desirable to consider aspects that relate the conservation of species and their habitats with the carbon storage outside forest areas. The extraction of biomass for the elaboration of long-lasting products such as furniture or infrastructure is a way to reduce the risk of leakage within this type of ecosystem, reducing the carbon content inside a warehouse, so that emissions risks are reduced and capture is encouraged by stimulating increases in forest mass in order reach the maximum biomass production potential.

5. Conclusions

The methodology used for the evaluation of a protected forest of *Abies religiosa* as aerial carbon storage and its capture capacity is adequate and reproducible in areas with similar conditions where it is not possible to use destructive methods. The use of high resolution satellite images combined with the analysis of physical aspects of the terrain allowed for detailed zoning directly related to the density of the forest mass, and its amount of biomass and carbon.

The results presented denote that the amount of carbon stored above-ground is directly related to the density and degree of disturbance. In this case, because it is a protected forest where the silvicultural activities are restricted, the productivity of the mass is lower, and consequently, the rate of carbon sequestration decreases.

The quantified carbon additionality is a parameter that depends on several factors, for which the age of the forest mass is considered one of the most important, and in the young and vigorous forests, the rate of sequestration is higher than in forests of advanced stages of development. In this study, *Abies religiosa* caught 1.03 MgC ha^{-1} yr^{-1}, with an annual average increase of 2.92%.

The implementation of silvicultural techniques governed by forest management with correct principles, foundations, and objectives greatly contributes to the reduction of atmospheric carbon. Within protected forests, the application of forest management techniques makes it possible to obtain sustainable forests and maximize the potential of forest soils to increase this important ecosystem service.

Acknowledgments: The authors thank the Mexican Government for providing the RapidEye satellite image used for stratification, as well as for the personnel who manage the El Chico National Park for the facilities granted to carry out this study.

Author Contributions: P.I.F.-L. and H.J.C.-B. drafted the document and elaborated, and designed the experiment; C.A.G.-R. contributed to the experimental design planning, data analysis, revision, and translation of the manuscript. E.M.O.-S. contributed to the revision of the document, and support and advice in the development of research. J.R.V.-L., R.R.-L., and R.R.-Z. were involved in document revision, data analysis, field phase support, and cartographic phase review.

Conflicts of Interest: The authors declare no conflict of interest.

References

1. Bassi, A.M.; Yudken, J.S.; Ruth, M. Climate policy impacts on the competitiveness of energy-intensive manufacturing sectors. *Energy Policy* **2009**, *37*, 3052–3060. [CrossRef]
2. De Vries, M.; de Boer, I. Comparing environmental impacts for livestock products: A review of life cycle assessments. *Livest. Sci.* **2010**, *128*, 1–11. [CrossRef]
3. Garnett, T. Livestock-related greenhouse gas emissions: Impacts and options for policy makers. *Environ. Sci. Policy* **2009**, *12*, 491–503. [CrossRef]
4. Getzner, M. The quantitative and qualitative impacts of clean technologies on employment. *J. Clean. Prod.* **2002**, *10*, 305–319. [CrossRef]
5. Gossling, S. *Carbon Management in Tourism: Mitigating the Impacts on Climate Change*; Routledge: Abingdon, UK, 2011.
6. Norgate, T.; Haque, N. Energy and greenhouse gas impacts of mining and mineral processing operations. *J. Clean. Prod.* **2010**, *18*, 266–274. [CrossRef]
7. Rivers, N. Impacts of climate policy on the competitiveness of Canadian industry: How big and how to mitigate? *Energy Econ.* **2010**, *32*, 1092–1104. [CrossRef]

8. Sano, F.; Akimoto, K.; Wada, K. Impacts of different diffusion scenarios for mitigation technology options and of model representations regarding renewables intermittency on evaluations of CO_2 emissions reductions. *Clim. Chang.* **2014**, *123*, 665–676. [CrossRef]

9. Valerio, F. Environmental impacts of post-consumer material managements: Recycling, biological treatments, incineration. *Waste Manag.* **2010**, *30*, 2354–2361. [CrossRef] [PubMed]

10. Bank, W. *Towards Sustainable Mineral-Intensive Growth in Orissa: Managing Environmental and Social Impacts*; World Bank: Washington, DC, USA, 2007; p. 78.

11. Brown, S. Present and Future Role of Forests in Global Climate Change. In *Ecology Today: An Anthology of Contemporary Ecological Research*; Goapl, B., Pathak, P.S., Saxena, K.G., Eds.; International Scientific Publications: New Delhi, India, 1998; pp. 59–74.

12. Metz, B.; Davidson, O.; Swart, R.; Pan, J. *Climate Change 2001: Mitigation*; Cambridge University Press: Cambridge, UK, 2001.

13. Watson, R.T.; Zinyowera, M.C.; Moss, R.H. *Climate Change 1995 Impacts, Adaptations and Mitigation of Climate Change: Scientific-Technical Analysis*; Cambridge University Press: Cambridge, UK, 1996.

14. Pan, Y.; Birdsey, R.A.; Fang, J.; Houghton, R.; Kauppi, P.E.; Kurz, W.A.; Phillips, O.L.; Shvidenko, A.; Lewis, S.L.; Canadell, J.G.; et al. A large and persistent carbon sink in the world's forests. *Science* **2011**, *333*, 988–993. [CrossRef] [PubMed]

15. Change, I.P.O.C. *Climate Change 2014–Impacts, Adaptation and Vulnerability: Regional Aspects*; Cambridge University Press: Cambridge, UK, 2014.

16. Nauclér, T.; Enkvist, P.-A. *Pathways to a Low-Carbon Economy: Version 2 of the Global Greenhouse Gas Abatement Cost Curve*; McKinsey & Company: New York, NY, USA, 2009.

17. Soares-Filho, B.; Moutinho, P.; Nepstad, D.; Anderson, A.; Rodrigues, H.; Garcia, R.; Dietasch, L.; Merry, F.; Bowman, M.; Hissa, L.; et al. Role of Brazilian Amazon protected areas in climate change mitigation. *Proc. Natl. Acad. Sci. USA* **2010**, *107*, 10821–10826. [CrossRef] [PubMed]

18. Venter, O.; Laurance, W.F.; Iwamura, T.; Wilson, K.A.; Fuller, R.A.; Possingham, H.P. Harnessing carbon payments to protect biodiversity. *Science* **2009**, *326*, 1368. [CrossRef] [PubMed]

19. Jackson, R.B.; Baker, J.S. Opportunities and constraints for forest climate mitigation. *BioScience* **2010**, *60*, 698–707. [CrossRef]

20. Lippke, B.; Perez-Garcia, J.; Manriquez, C. *Executive Summary: The Impact of Forests and Forest Management on Carbon Storage, Rural Technological Initiative, College of Forest Resources*; Box: Redwood City, CA, USA, 2003.

21. Food and Agriculture Organization of the United Nations (FAO). *Global Forest Resources Assessment 2015*; Organización de las Naciones Unidas Para la Alimentación y la Agricultura: Rome, Italy, 2015.

22. Oates, J.F. *Myth and Reality in the Rain Forest: How Conservation Strategies Are Failing in West Africa*; University of California Press: Berkeley, CA, USA, 1999.

23. Ferraro, P.J.; Pattanayak, S.K. Money for nothing? A call for empirical evaluation of biodiversity conservation investments. *PLoS Biol.* **2006**, *4*, e105. [CrossRef] [PubMed]

24. Food and Agriculture Organization of the United Nations (FAO). *Global Forest Resources Assessment 2015; How Are the World's Forests Changing?* Food and Agriculture Organization of the United Nations: Rome, Italy, 2016.

25. Leverington, F.; Hockings, M.; Costa, K.L. *Management Effectiveness Evaluation in Protected Areas—A Global Study*; Supplementary Report No. 1. Overview of Approaches and Methodologies; World Commission on Protected Areas: Brisbane, Australia, 2008.

26. Nolte, C.; Leverington, F.; Kettner, A.; Marr, M.; Nielsen, G.; Bomhard, B.; Stolton, S.; Stoll-Kleemann, S.; Hockings, M.; World Conservation Union (IUCN)-World Commission on Protected Areas (WCPA). *Protected Area Management Effectiveness Assessments in Europe*; A Review of Application, Methods and Results; BfN-Skripten: Bonn, Germany, 2010; p. 69.

27. Zabala, F. Análisis demográfico preliminar de Taxus globosa Schlecht en el Parque Nacional El Chico, Hidalgo, Mexico. I: Población de adultos y algunas características del hábitat. *CIENCIA Ergo-Sum* **2001**, *8*, 169–174.

28. Razo Zárate, R.; Martínez, A.J.G.; Laguna, R.R.; Maycotte Morales, C.C.; Acevedo Sandoval, O.A. Coeficientes de carbono para arbustos y herbáceas del bosque de oyamel del Parque Nacional El Chico. *Rev. Mex. Cienc. For.* **2016**, *6*, 10.

29. Melo Gallegos, C.; López García, J. Parque Nacional El Chico, marco geográfico-natural y propuesta de zonificación para su manejo operativo. *Investig. Geogr.* **1994**, *28*, 65–128. [CrossRef]

30. Cervantes, Á.; Reyes, E. Efectos Ecológicos de los Incendios Forestales Sobre el Bosque de Oyamel. Ph.D. Thesis, Colegio de Postgraduados Campus Montecillo, Texcoco, Mexico, 2010.

31. Leopold, A.S. Vegetation zones of Mexico. *Ecology* **1950**, *31*, 507–518. [CrossRef]

32. Rzedowski, J. *Vegetación de Mexico. 1ra*; Edición Digital; Comisión Nacional Para el Conocimiento y Uso de la Biodiversidad: Mexico City, Mexico, 2006; p. 504.

33. Castellanos Bolaños, J.F.; Gómez Cárdenas, M.; Contreras Hinojosa, J.R.; González Cubas, R. *Metodologías Para Cuantificar Biomasa y Carbono en Bosques*; Centro de Investigación Regional del Pacífico Sur: Oaxaca, Mexico, 2013.

34. Návar-Cháidez, J.D.J.; Domínguez-Calleros, P.A. Modelo de incremento y rendimietno: Ejemplos y aplicaciones para bosques templados mexicanos. *Rev. Mex. Cienc. For.* **2013**, *4*, 8–27.

35. Rafael Domingo, C.M. Monitoreo de bosques utilizando ndvi rededge de rapideye. *Instr. Gener. Para Autores* **2013**, *10*, 58–71.

36. Aguilar Arias, H.; Zamora, R.M.; Bolaños, C.V. Metodología Para La Corrección Atmosférica De Imágenes Aster, Rapideye, Spot 2 Y Landsat 8 Con El Módulo Flaash Del Software Envi. Atmospheric Correction Methodology for Aster, Rapideye, Spot 2 and Landsat 8 Images with Envi Flaash Module Software. *Rev. Geogr. Am. Cent.* **2015**, *2*. [CrossRef]

37. Gitelson, A.; Merzlyak, M.N. Spectral Reflectance Changes Associated with Autumn Senescence of Aesculus *hippocastanum* L. and *Acer platanoides* L. Leaves. Spectral Features and Relation to Chlorophyll Estimation. *J. Plant Physiol.* **1994**, *143*, 286–292. [CrossRef]

38. Sims, D.A.; Gamon, J.A. Relationships between leaf pigment content and spectral reflectance across a wide range of species, leaf structures and developmental stages. *Remote Sens. Environ.* **2002**, *81*, 337–354. [CrossRef]

39. Tapsall, B.; Milenov, P.; Tasdemir, K. Analysis of RapidEye Imagery for Annual landcover Mapping As an Aid to European Union (EU) Common Agricultural Policy. In Proceedings of the International Archives of the Photogrammetry, Remote Sensing and Spatial Information Sciences, Vienna, Austria, 5–7 July 2010; pp. 568–573.

40. Wu, C.; Niu, Z.; Tang, Q.; Huang, W.; Rivaed, B.; Feng, J. Remote estimation of gross primary production in wheat using chlorophyll-related vegetation indices. *Agric. For. Meteorol.* **2009**, *149*, 1015–1021. [CrossRef]

41. Roslani, M.A.; Mustapha, M.A.; Lihan, T.; Wan Juliana, W.A. Applicability of Rapideye Satellite Imagery in Mapping Mangrove Vegetation Species at Matang Mangrove Forest Reserve, Perak, Malaysia. *J. Environ. Sci. Technol.* **2014**, *7*, 123.

42. Aguirre, O.; Jiménez, J.; Treviño, E.; Meraz, B. Evaluación de diversos tamaños de sitio de muestreo en inventarios forestales. *Madera Bosques* **1997**, *3*, 71–79. [CrossRef]

43. Juárez Castillo, S. *Trial to Determine Comparative Efficiency of Sampling Sites in Temperate-and Cold-Climate Forests*; Centro Interamericano de Fotointerpretación: Bogotá, Colombia, 1974.

44. Fedrigo, M.; Kasel, S.; Bennett, L.T.; Roxburgh, S.H.; Nitschke, C.R. Carbon stocks in temperate forests of south-eastern Australia reflect large tree distribution and edaphic conditions. *For. Ecol. Manag.* **2014**, *334*, 129–143. [CrossRef]

45. Klepac, D. *Crecimiento e Incremento de Árboles y Masas Forestales*; Universidad Autónoma Chapingo: Texcoco, Mexico, 1976.

46. Muñoz-Ruiz, M.Á.; Valdez-Lazalde, J.R.; de los Santos-Posadas, H.M.; Ángeles-Pérez, G.; Monterroso-Rivas, A.I. Inventario y mapeo del bosque templado de Hidalgo, Mexico mediante datos del satélite SPOT y de campo. *Agrociencia* **2014**, *48*, 847–862.

47. Hernández-Díaz, J.C.; Corral-Rivas, J.J.; Quiñones-Chávez, A.; Bacon-Sobbe, J.R.; Vargas-Larreta, B. Evaluación del manejo forestal regular e irregular en bosques de la Sierra Madre Occidental. *Madera Bosques* **2008**, *14*, 25–41. [CrossRef]

48. Razo-Zárate, R.; Gordillo-Martínez, A.J.; Rodríguez-Laguna, R.; Maycotte-Morales, C.C.; Acevedo-Sandoval, O.A. Estimación de biomasa y carbono almacenado en árboles de oyamel afectados por el fuego en el Parque nacional "El Chico", Hidalgo, Mexico. *Madera Bosques* **2013**, *19*, 73–86. [CrossRef]

49. Villa Salas, A.B.; Aguilar Ramírez, M. Rutinas de cálculo de once métodos para determinar el incremento en volumen de conífertas. *Rev. Mex. Cienc. For.* **2012**, *20*, 77.

50. Avendaño Hernandez, D.M.; Mireles, M.A.; Anzures, F.C.; Etchevers Barra, J.D. Estimación de Biomasa y Carbono en un Bosque de Abies Religiosa. *Rev. Fitotec. Mex.* **2009**, *32*, 233–238.

51. Leñero, L.A.; Nava, M.; Ramos, A.; Espinosa, M.; De Jesus Ordonez, M.; Jujnovsky, J. Servicios ecosistémicos en la cuenca del río Magdalena, Distrito Federal, Mexico. *Gaceta Ecol.* **2007**, *84–85*, 53–64.
52. Ávila-Akerberg, V. *Forest Quality in the Southwest of Mexico City: Assessment towards Ecological Restoration of Ecosystem Services*; Institut für Landespflege: Freiburg im Breisgau, Germany, 2010.
53. Razo-Zárate, R.; Gordillo-Martínez, A.; Rodríguez-Laguna, R.; Maycotte-Morales, C.; Acevedo-Sandoval, O. Escenarios de carbono para el bosque de oyamel del Parque Nacional El Chico, Hidalgo, Mexico. *Rev. Latinoam. Rec. Nat.* **2013**, *9*, 17–21.
54. Mendoza-Ponce, A.; Galicia, L. Aboveground and belowground biomass and carbon pools in highland temperate forest landscape in Central Mexico. *Forestry* **2010**, *83*, 497–506. [CrossRef]
55. Maass, S.F. *Estimación de la Captura de Carbono en Zonas Forestales: El Caso del Parque Nacional Nevado de Toluca*; Universidad Autónoma del Estado de Mexico: Toluca, Mexico, 2009.
56. Pregitzer, K.S.; Euskirchen, E.S. Carbon cycling and storage in world forests: Biome patterns related to forest age. *Glob. Chang. Biol.* **2004**, *10*, 2052–2077. [CrossRef]
57. Aguirre-Salado, C.A.; Valdez-Lazalde, J.R.; Ángeles-Pérez, G.; de los Santos-Posadas, H.M.; Haapanen, R.; Aguirre-Salado, A.I. Mapeo de carbono arbóreo aéreo en bosques manejados de pino Patula en Hidalgo, Mexico. *Agrociencia* **2009**, *43*, 209–220.
58. Masera, O.R.; Cerón, A.D.; Ordóñez, A. Forestry mitigation options for Mexico: Finding synergies between national sustainable development priorities and global concerns. *Mitig. Adapt. Strateg. Glob. Chang.* **2001**, *6*, 291–312. [CrossRef]
59. Huntington, T.G. Carbon Sequestration in an Aggrading Forest Ecosystem in the Southeastern USA. *Soil Sci. Soc. Am. J.* **1995**, *59*, 1459–1467. [CrossRef]
60. Rojo, J.M.T.; Sanginés, A.G. El potencial de Mexico para la producción de servicios ambientales: Captura de carbono y desempeño hidráulico. *Gac. Ecol.* **2002**, *63*, 40–59.
61. Bonan, G.B. Forests and Climate Change: Forcings, Feedbacks, and the Climate Benefits of Forests. *Science* **2008**, *320*, 1444–1449. [CrossRef] [PubMed]
62. Canadell, J.G.; Raupach, M.R. Managing Forests for Climate Change Mitigation. *Science* **2008**, *320*, 1456–1457. [CrossRef] [PubMed]
63. Seely, B.; Welham, C.; Kimmins, H. Carbon sequestration in a boreal forest ecosystem: Results from the ecosystem simulation model, FORECAST. *For. Ecol. Manag.* **2002**, *169*, 123–135. [CrossRef]
64. Chacón, P.; Leblanc, H.; Russo, R. Fijación de carbono en un bosque secundario de la región tropical húmeda de Costa Rica. *Tierra Trop.* **2007**, *3*, 1–11.
65. Figueroa-Navarro, C.M.; Ángeles-Pérez, G.; Velázquez-Martínez, A.; de los Santos-Posadas, H.M. Estimación de la biomasa en un bosque bajo manejo de Pinus patula Schltdl. et Cham. en Zacualtipán, Hidalgo. *Rev. Mex. Cienc. For.* **2010**, *1*, 105–112.
66. Harmon, M.E.; Harmon, J.M.; Ferrell, W.K.; Brooks, D. Modeling carbon stores in Oregon and Washington forest products: 1900–1992. *Clim. Chang.* **1996**, *33*, 521–550. [CrossRef]
67. CCMSS. *El Manejo Sostenible de los Bosques Como Estrategia de Combate al Cambio Climático en México*; Punto Verde Consultores, S.C.: Monterrey, Mexico, 2010.
68. Liu, J.; Liu, S.; Loveland, T.R. Temporal evolution of carbon budgets of the Appalachian forests in the US from 1972 to 2000. *For. Ecol. Manag.* **2006**, *222*, 191–201. [CrossRef]
69. Zhang, J.; Ge, Y.; Chang, J.; Jiang, B.; Jiang, H.; Peng, C.; Zhu, J.; Yuan, W.; Qi, L.; Yu, S. Carbon storage by ecological service forests in Zhejiang Province, subtropical China. *For. Ecol. Manag.* **2007**, *245*, 64–75. [CrossRef]
70. Manzanilla, H. *Investigaciones Epidométricas y Silvícolas en Bosques Mexicanos de Abies Religiosa*; Secretaría de Agricultura y Ganadería, Dirección General de Información y Relaciones Públicas: Mexico City, Mexico, 1974.

 *forests*

Article

Nonlinear Variations of Net Primary Productivity and Its Relationship with Climate and Vegetation Phenology, China

Jian Yang [1,2], Xin Chang Zhang [1,*], Zhao Hui Luo [2] and Xi Jun Yu [2]

[1] School of Geography and Planning, Sun Yat-sen University, Guangzhou 510275, China; yangjian@scies.org
[2] South China Institute of Environmental Sciences, Ministry of Environmental Protection, Guangzhou 510655, China; luozhaohui@scies.org (Z.H.L.); yuxijun@scies.org (X.J.Y.)
* Correspondence: eeszxc@mail.sysu.edu.cn; Tel.: +86-020-8411-5833

Received: 8 August 2017; Accepted: 19 September 2017; Published: 25 September 2017

Abstract: Net primary productivity (NPP) is an important component of the terrestrial carbon cycle. In this study, NPP was estimated based on two models and Moderate Resolution Imaging Spaectroradiometer (MODIS) data. The spatiotemporal patterns of NPP and the correlations with climate factors and vegetation phenology were then analyzed. Our results showed that NPP derived from MODIS performed well in China. Spatially, NPP decreased from the southeast toward the northwest. Temporally, NPP showed a nonlinear increasing trend at a national scale, but the magnitude became slow after 2004. At a regional scale, NPP in Northern China and the Tibetan Plateau showed a nonlinear increasing trend, while the NPP decreased in most areas of Southern China. The decreases in NPP were more than offset by the increases. At the biome level, all vegetation types displayed an increasing trend, except for shrub and evergreen broad forests (EBF). Moreover, a turning point year occurred for all vegetation types, except for EBF. Generally, climatic factors and Length of Season were all positively correlated with the NPP, while the relationships were much more diverse at a regional level. The direct effect of solar radiation on the NPP was larger (0.31) than precipitation (0.25) and temperature (0.07). Our results indicated that China could mitigate climate warming at a regional and/or global scale to some extent during the time period of 2001–2014.

Keywords: net primary production; spatiotemporal patterns; climate change; phenology; China

1. Introduction

Net primary productivity (NPP) is the net amount of carbon accumulated by plants in a given period, and has been regarded as one of the main components of the carbon cycle [1]. Due to a variety of direct and/or indirect anthropogenic activities (e.g., land clearing and conversion) and nature disturbances (e.g., fire, pests) as well as global and regional climate change, forested ecosystems have undergone substantial changes in cover and have increasingly shown declines in health over recent decades [2,3]. As a sensitive indicator of forest cover, function and health, NPP loss may affect the composition of the atmosphere, fresh water availability, biodiversity [4,5], and the ecological adjusting mechanism of energy supply and distribution [6].

Quantifying the inter-annual variability in NPP and the interactions with climate factors would help us understand the terrestrial carbon dynamics and underlying mechanisms in responses to climate change [7]. Recently, many studies have been undertaken on the spatiotemporal variation of NPP and its relationship with climate factors on global and regional scales using a range of approaches from observational [2,8,9] to a suite of remote sensing based methods [10,11]. However, the NPP results were diverse in trends and magnitudes, even in the same region among models [10,12] due to different inputs. Moderate Resolution Imaging Spaectroradiometer (MODIS) annual NPP products have been widely used [13,14]; however, uncertainties from inputs and the algorithm resulted in biases and restrained data use regionally or globally to some extent. For instance, 70–80% accuracy of MODIS land cover products (MOD12Q1) are an assimilated meteorological dataset, not observed data with coarse spatial resolution, the cloud-contaminated MODIS FPAR/LAI (MOD15A2), and weaknesses in the MOD17 algorithm [15,16]. Therefore, it is essential to compare MODIS derived NPP to other models and/or field observed NPP, especially for China, which encompasses a wide range of ecosystems and climates. In addition, the responses of different ecosystems to different magnitudes of climate change are still far from clear, especially in China [17]. Moreover, few studies [12] have investigated the effect of solar radiation as an integrated surrogate for the effects of both day length and sunlight intensity [18] on vegetation NPP, especially when temperature and precipitation are considered simultaneously. In addition, previous studies have shown that vegetation phenology, an important factor that affects plant productivity, has changed dramatically due to climate changes and anthropogenic interference [19], but only a few studies have explored the effect of phenophase variation on NPP [20]. At present, due to the diversity of the trends and magnitudes of vegetation phenophases at different scales, its effects on NPP are still unclear, especially in China, where the vegetation phenophases are diverse at both the regional and biome levels [19]. Furthermore, a linear regression method has been applied by most studies to analyze NPP trend [10,21] despite the trend always showing a non-linear trend, or has one or more turning points within the time period [22,23]. These limitations and/or gaps have impeded our understanding of the dynamic relationship and consequently researchers may have underestimated future changes in plant productivity and/or the carbon cycle throughout China.

China encompasses a variety of ecosystems and climates. The regional climate ranging from tropical to cold-temperate, and from humid in the south to extremely dry in the northwest [24]. Land cover types are diverse, including a broad range of tropical, temperate and boreal forests, grassland, cropland and desert [25]. In recent decades, China has experienced dramatic changes in climate such as remarkably strong El Niño events [26], the freezing low temperatures in early 2008 [24], and frequent occurrences of severe droughts [27]. Meanwhile, land use and land cover changes have occurred at unprecedented rates due to quick economic development, dramatic urbanization [28], and implementation of several large scale forest plantation programs [29]. These changes have resulted in large variations in China's terrestrial ecosystem productivity and have definitely adjusted the terrestrial carbon cycle in China [30]. However, whether the temporal trend of NPP continuously increased or decreased during the study period is still unclear given the possible changes of climate derivers and anthropogenic activities. Additionally, the correlations between NPP and climate derivers as well as vegetation phenophases in recent decade still remain unclear [31].

Due to climatic variability, topographic complexity, natural ecosystem diversity, and intensive human disturbance, China is becoming one of the most critical and sensitive regions in the global carbon cycle for determining the carbon budget at regional and global scales. Furthermore, it provides a good opportunity to identify the effects of climate change on NPP to forecast the potential biosphere feedback to nature in the climate system.

Therefore, two simple models were applied to estimate NPP and the results accompanied by MODIS derived NPP were all compared with the field observed NPP. Then, the results that showed less biases with the field observed NPP were selected to analyze the vegetation NPP dynamics and its relationship with both climate and vegetation phenology in China. More specifically, the authors aim was to (1) explore which model's result was more accurate, and the quantity of uncertainty of MODIS derived NPP in China; (2) understand the spatial pattern of NPP, and investigate whether the temporal trend of NPP was continuously increasing or decreasing in China for the period 2001–2014 given climate change and anthropogenic activities; and (3) estimate the effects of climatic driving factors and vegetation phenology changes on vegetation NPP.

2. Materials and Methods

2.1. Study Area

The study focused only on Mainland China. The climate in China is extremely diverse, ranging from tropical regions in the south to subarctic in the north [11]. Therefore, the whole study area was divided into three sub-regions (Figure S1) according to the climatic regionalization of China [32]: (1) Northern China, with mean annual temperature ranges from $-4\ ^\circ$C to $14\ ^\circ$C and total annual precipitation ranges from 200 mm in the northwest to 1000 mm in the southeast; (2) Southern China, where mean annual temperature ranges from $14\ ^\circ$C to $22\ ^\circ$C and total annual precipitation ranges from 1000 to 2000 mm [33]; and (3) the Tibetan Plateau, which has been called the Third Pole of the World, with an average altitude close to 4000 m above sea level, a mean annual temperature ranging from $-5\ ^\circ$C to $12\ ^\circ$C and precipitation ranging from >800 mm to <200 mm [34].

2.2. Dataset and Data Processing

2.2.1. Annual Net Primary Productivity (NPP)

NPP derived from MODIS from 2001–2014 in China were used (MOD17A3, 1 km), which have been further used as an important data source for plant productivity monitoring and assessment [35]. The MODIS annual NPP algorithm relies on the summation of the daily estimation of Gross Primary Productivity (GPP) computed globally minus growth and maintenance respiration [14]. To estimate annual NPP, first, the daily estimates of maintenance respiration for leaves and fine roots are subtracted from the daily GPP values. These daily reduced GPP estimates are then summed for each year, and estimates for annual maintenance respiration of living wood tissue and annual total growth respiration are subsequently subtracted, resulting in annual NPP estimates. Full details of the algorithm in MODIS derived NPP can be found in Reference [14]. However, due to the deficiencies and uncertainty of MODIS derived NPP as above-mentioned, the Miami model and Thornthwaite Memorial model were applied in this study, as they are simple to operate with only a few parameters that were easy to obtain, and have a higher spatial and temporal resolution of climate data in China. Additionally, they have previously been applied in China [11,13]. The Miami model is calculated as follows (Equations (1)–(3)):

$$NPP_{T,P} = \min\{f_1(T),\ f_2(P)\} \tag{1}$$

$$f_1(T) = \frac{k_1}{1 + e^{k_2 - k_3 \times T}} \tag{2}$$

$$f_1(P) = k_4 \times \left(1 - e^{-k_5 \times P}\right) \tag{3}$$

where K_1, K_2 and K_3 in Equation (2) are the temperature response parameters with values of 3000, 1.315 and 0.119, respectively and T is the annual average temperature ($^\circ$C); K_4 and K_5 in Equation (3) are the precipitation response parameters with values of 3000 and 0.000664, respectively, and P is the annual average precipitation (mm). The parameters of K_i ($i = 1, 2, \ldots, 5$) were calculated using the least squares method based on the field testing of NPP, and the relative temperature and precipitation collected at 50 locations scattered across five continents [36].

The Thornthwaite Memorial model is expressed as follows (Equations (4)–(6)):

$$NPP = 3000 \times \left(1 - e^{-0.0009695 \times |v - 20|}\right) \tag{4}$$

$$v = \frac{1.05 \times R}{\sqrt{1 + (1 + 1.05\frac{R}{L})^2}} \tag{5}$$

$$L = 3000 + 25 \times T + 25T^3 \tag{6}$$

where v represents the mean annual actual evapotranspiration (mm); L represents the mean annual evapotranspiration (mm); and T and R represent the mean annual temperature and mean annual precipitation, respectively.

Due to the different inputs and parameters in the three models, all the estimated NPP results were compared with the field observed NPP (Table S1), which was calculated from flux tower data based on the eddy covariance method, a micrometeorological method with high-frequency data collection. This method provides direct measures of net carbon fluxes between vegetated canopies and the atmosphere over short and long timescales with minimal disturbance to the underlying vegetation. The observation data were collected from ChinaFLUX sites [37] directly and indirectly from the literature. High-precision CO_2 concentration measurements and surface carbon flux measures, made at eddy covariance measurement sites, can be used to improve and validate the algorithms being used by remote sensing and ecosystem models [38]. The observations almost covered the typical ecosystem types in China. Therefore, observations from carbon flux data were applied as a reference to compare the results from the three models. Sites only having one year, the corresponding yearly data for observations, and three simulated values were used directly; and for sites longer than one year, the corresponding multi-year averaged data were applied. Next, two statistical indicators, namely root mean square error (RMSE), and mean absolute error (MAE), were used to evaluate the performance of each method above-mentioned. The best estimated result was then considered in the following analysis of the NPP relationship with climatic factors and vegetation phenology. The two statistical indicators were calculated as follows (Equations (7) and (8)):

$$\text{RMSE} = \sqrt{\frac{\sum_{i=1}^{n}(NPP_i - NPP^*)}{n}} \tag{7}$$

$$\text{MAE} = \frac{1}{n}\sum_{i=1}^{n}|NPP_i - NPP^*| \tag{8}$$

where NPP_i is the estimated NPP, NPP^* is the field observed NPP; and n is the sample size.

2.2.2. Climate Data and Phenology Extraction

Monthly meteorological data from 2001–2014, including temperature, precipitation, and solar radiation, were acquired from the China Meteorological Data Sharing Service System (downloaded from [39]) All meteorological data used in this study were verified by China's Meteorological Information Center (located in Beijing, China) [40], thus false or missing data from some of the stations were eliminated [11]. There are 653 meteorological stations recording temperature and precipitation data (301 in Northern China, 258 in Southern China, and 94 in the Tibetan Plateau), and 99 stations recording solar radiation (50 in Northern China, 38 in Southern China, and 11 in the Tibetan Plateau; Figure S2). The Kriging method was used for the spatial interpolation of climate data across the study area [41].

Vegetation phenology metrics (Start of Season (SOS) and End of Season (EOS) for each year) were extracted using TIMESAT software (which is widely used for simulating vegetation phenology [42,43]) by applying the Savitzky–Golay (S–G) method to generate smooth time-series MODIS EVI data. We adopted an adaptation strength of 2.0, no spike filtering, seasonal parameter of 0.5, S–G window size of 2, and amplitude season start and end of 20% to calculate the phenology parameters. Length of Season (LOS) was calculated as the difference between the SOS and EOS values.

2.2.3. Land Cover Data

A 1-km spatial resolution land cover product, MOD12Q1, was applied in this study to analyze NPP and its variation across vegetation types. The MOD12Q1 land cover dataset which was extensively applied to monitor land use and land cover change [44], is mainly based on the International Geosphere–Biosphere program (IGBP) classification system that obtains a classification algorithm of the decision tree and artificial neural network [45]. The MOD12Q1 land cover classes were further reclassified into seven major land cover classes (Figure S1) in the present study: evergreen needle-leaf forest (ENF), evergreen broadleaf forest (EBF), deciduous needle-leaf forest (DNF), deciduous broadleaf forest (DBF), farmland grassland and meadow (GM), and shrubs. However, the data only reflected the land cover classifications and could not consider changes in land cover that occurred over our study period. It was estimated that vegetation changes might be relatively stable over a short time period of approximately 10 years at a regional or global scale [42].

2.3. Data Processing

2.3.1. Trend Analysis and Turning Point Year Detection

The trends of NPP during 2001–2014 were calculated at both pixel level and regional level. Due to autocorrelation among the inter-annual time series data, a robust non-parametric Mann–Kendall (M–K) trend analysis [46] was applied. This method did not require the independence and normality of the time series data [47], which has been widely used in trend analysis [19]. Previous studies have reported that the M–K test statistic Z was approximately normally distributed when the sample size was $n \geq 8$. A positive or a negative Z value indicated an increasing or a decreasing trend respectively, which were all monotonic [46]. The formulas for the M–K method are described in detail in Reference [46]. In addition, trends of the NPP were tested at a significance level of $\alpha = 0.05$.

The Theil–Sen median slope estimator was applied to estimate the rate of change of NPP, which was more appropriate for assessing the rate of change in short or noisy time series [48]. The Theil–Sen median slope was computed as (Equation (9)):

$$\beta_i = \text{Median}(\frac{x_j - x_k}{j - k}) \text{ for } i = 1, \ldots, N \tag{9}$$

where x_j and x_k are the data values at times j and k ($j > k$), respectively.

To detect the timing and magnitude of NPP changes, a linear regression model and a piecewise linear regression model (Equation (10)) were used. This method has been widely used [41,49] as it can detect potential turning points (TPs) in a trend of time-series data. In addition, TP was limited to the years 2004 to 2010 to avoid obtaining a too-short segment before or after the TP [50]. The maximum number of TPs was specified as one given that too many TPs would make the result more complex and thus create more uncertainty in the understanding of NPP trends [23]. The models used were:

$$Y = \begin{cases} \beta_1 \times t + \beta_0 + \varepsilon & (t \leq \alpha) \\ \beta_1 \times t + \beta_2 \times (t - \alpha) + \beta_0 + \varepsilon & (t > \alpha) \end{cases} \tag{10}$$

where Y is the NPP; t is the year; α is the turning point of the NPP time series; β_0 is the intercept; β_1 is the magnitude of the NPP trend before the TP; $(\beta_1 + \beta_2)$ is the magnitude of the NPP trend after the TP; α is the year of the TP; and ε is the residual random error.

The two models were then fitted to the NPP time-series data using the least-squares method, and the Akaike Information Criterion (AIC) [51] was used to determine whether the TPs were significant as it provides a means for model selection [41,51]. The AIC values of these two models were calculated as (Equation (11)):

$$AIC = n \times \log\left(\frac{RSS}{n}\right) + 2k + \frac{2k(k + 1)}{n - k - 1} \tag{11}$$

where RSS is the residual sum of squares for the estimated model; k is the number of parameters; and n is the sample size.

Finally, ΔAIC was defined as the difference of the AIC2 of the piecewise linear regression model (model 2) and the AIC1 of linear regression model (model 1); if the ΔAIC was less than -2, then model 2 was significantly preferred [8].

2.3.2. Correlations Relating Climatic Factors, LOS and NPP

Partial correlation analysis was used to explore the relationships between NPP and climatic factors (mean annual temperature, annual cumulative precipitation, and annual cumulative solar radiation from 2001–2014) as well as LOS, which excluded the effects of other variables. The significance of the correlation coefficients was tested at a significance level of 0.05.

To understand the direct and indirect effects of climatic factors, LOS on NPP variation (excluding multi-collinearity between climatic factors and LOS), a structural equation model (SEM) with standardized data was applied to examine the influences of climate and LOS change on NPP variation. The SEM was initiated by including all possible relationships. The least significant relationship was then removed stepwise until all relationships were significant and the fit of the model did not increase further [52]. SEM was conducted by the "sem" R package [53], which was then visualized by the R package "semPlot".

3. Results

3.1. Accuracy Assessment of NPP Estimation

The accuracy of three modeled NPP results (MODIS derived, the Miami model, and the Thornthwaite Memorial model) were assessed by a comparison with the in situ observations. Validation analysis indicated that the modeled NPP results were all very significantly (Figure 1; $p < 0.01$) correlated with the in situ NPP measurements. The R^2 value for the MODIS derived NPP was 0.39, which was higher than that of Miami model ($R^2 = 0.27$) and the Thornthwaite Memorial model ($R^2 = 0.28$). Furthermore, the RMSE and MAE deceased to 0.19 kg C m^{-2} a^{-1} and 0.14 kg C m^{-2} a^{-1}, respectively for the MODIS NPP data compared to 0.29 kg C m^{-2} a^{-1} and 0.23 kg C m^{-2} a^{-1} for the Miami model, as well as 0.31 kg C m^{-2} a^{-1} and 0.26 kg C m^{-2} a^{-1} for the Thornthwaite Memorial model, respectively. The results demonstrated that the NPP derived from MODIS had an improved performance of NPP simulation by increasing R^2 by 46.18% and 40.46% when compared with the Miami model and Thornthwaite Memorial models, respectively. In addition, the NPP derived from MODIS decreased the relative RMSE and MAE by 34.01% and 38.34%, respectively, when compared with the Miami model, and by 39.71% and 45.88%, respectively, when compared with the Thornthwaite Memorial model.

3.2. Spatial Pattern of NPP

The spatial pattern of mean annual NPP derived from MODIS for the period 2001–2014 is shown in Figure 2a. The spatial pattern of annual NPP was uneven, which showed gradients decreasing from the south to the north and from the east to the west. The highest value occurred in Southern China, with values generally higher than 0.6 kg C m^{-2} in most of that area. In contrast, annual NPP was usually lower than 0.2 kg C m^{-2} in the Tibetan Plateau. For the remaining regions, annual NPP ranged between 0.2 and 0.6 kg C m^{-2}. The standard deviation (Figure 2b) shows a similar spatial pattern with mean annual NPP. The highest value (more than 0.05 kg C m^{-2}) was mostly distributed in Southern China. The values in Northern China ranged from 0.02–0.05 kg C m^{-2}, with the exception of the northern and eastern parts of Northeast China, which was more than 0.05 kg C m^{-2}. The lowest standard deviation value was found in the Tibetan Plateau, with values mostly less than 0.02 kg C m^{-2}.

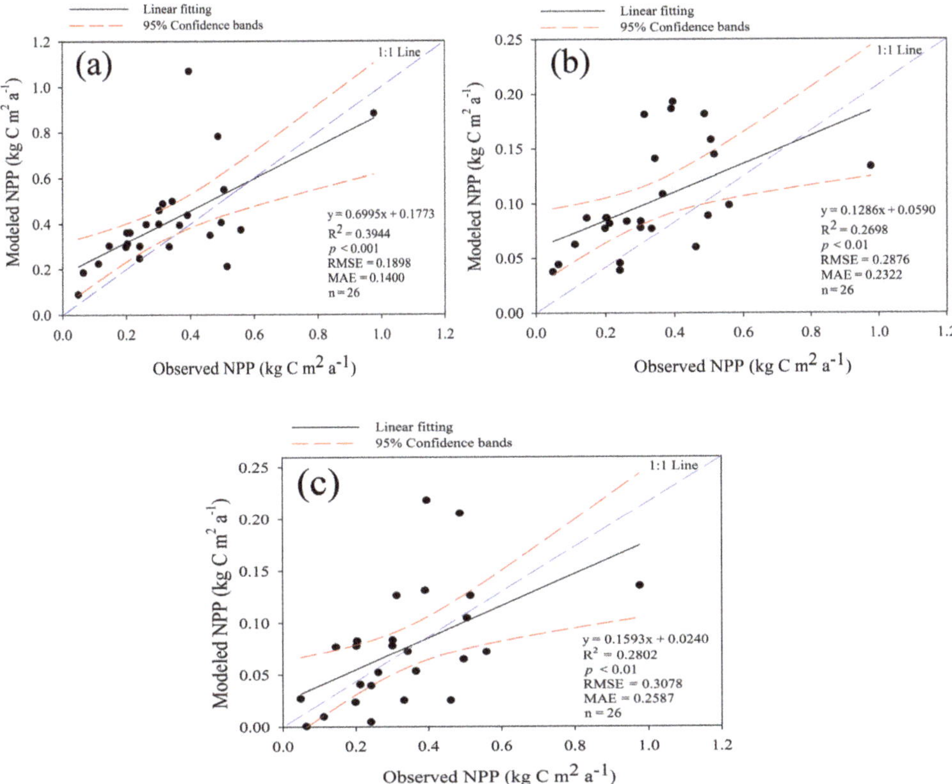

Figure 1. Comparison between the in situ Net Primary Productivity (NPP) site observations and NPP derived from MODIS (**a**); Miami model (**b**); and Thornthwaite Memorial model (**c**).

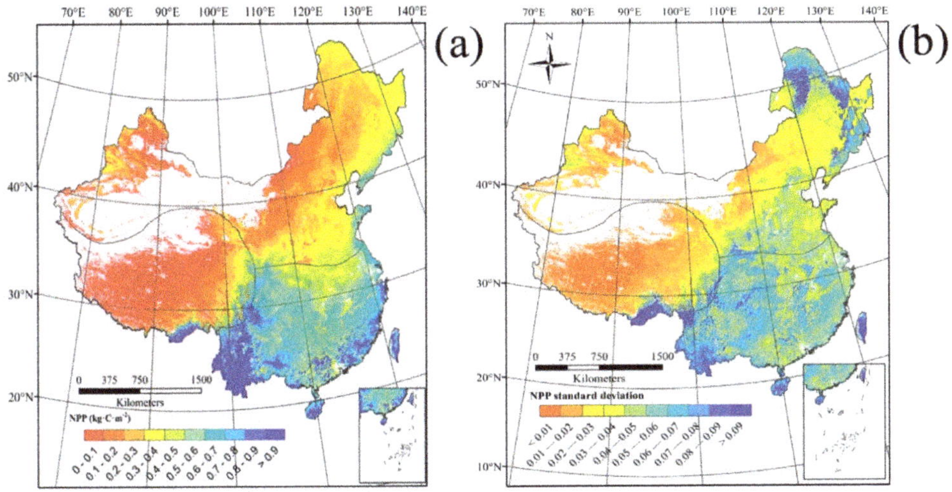

Figure 2. Spatial distribution of mean Net Primary Productivity (NPP) (**a**) and standard deviation (**b**) in China between 2001–2014.

At the biome level, NPP differs in terms of vegetation types (Figure 3). Generally, forest ecosystems have a higher annual NPP than grassland ($p < 0.001$). Among the forest ecosystems, evergreen forests show a higher NPP than deciduous forests ($p < 0.001$). The highest NPP was found in EBF, with an average annual value of 0.975 ± 0.362 kg C m^{-2}, which was almost twice that of the DBF (0.542 ± 0.301 kg C m^{-2}) and DNF (0.329 ± 0.103 kg C m^{-2}). ENF had the second largest NPP (0.605 ± 0.277 kg C m^{-2}), followed by Farmland (0.422 ± 0.159 kg C m^{-2}) and Shrub (0.317 ± 0.273 kg C m^{-2}), respectively. Grassland had the lowest annual NPP, with an average value of 0.162 ± 0.196 kg C m^{-2}.

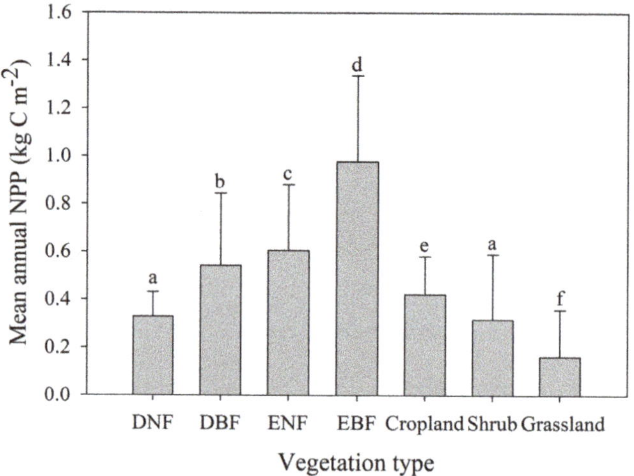

Figure 3. Mean annual Net Primary Productivity (NPP) for each vegetation type over 2001–2014. The error bar in each column indicates the standard deviation. Different lowercase letters indicate significant differences in NPP among vegetation types (one-way ANOVA and *t*-test, $p < 0.001$).

3.3. Temporal Trend of NPP

Figure 4 details the temporal trends in annual NPP across China. Over a 14-year period (2001–2014), 17.69% of total pixels displayed either significantly increased or decreased trends (Figure 4b; $p < 0.05$) in the whole study region. Pixels with a positive (i.e., increased NPP) trend in NPP accounted for 78.94% of the total pixels, and 15.18% of the total pixels exhibited a significantly positive trend (Figure 4b,c; $p < 0.05$). The increasing rate of NPP in China mainly occurred between 0 and 0.4×10^{-2} kg C m^{-2} a^{-1} (Figure 4c). At a regional scale, 79.15% of the total pixels showed a positive trend in Northern China, and the increasing rate was mostly distributed in 0–0.6×10^{-2} kg C m^{-2} a^{-1} (Figure 4d). Similarly, 88.12% of the total pixels displayed an increasing trend in the Tibetan Plateau, where an increasing magnitude mostly occurred at 0–0.4×10^{-2} kg C m^{-2} a^{-1} (Figure 4f). In contrast, negative trends occurred in 57.64% of the total pixels in Southern China, with an absolute magnitude mainly ranging from 0.2×10^{-2} kg C m^{-2} a^{-1} to 0.8×10^{-2} kg C m^{-2} a^{-1} (Figure 4e).

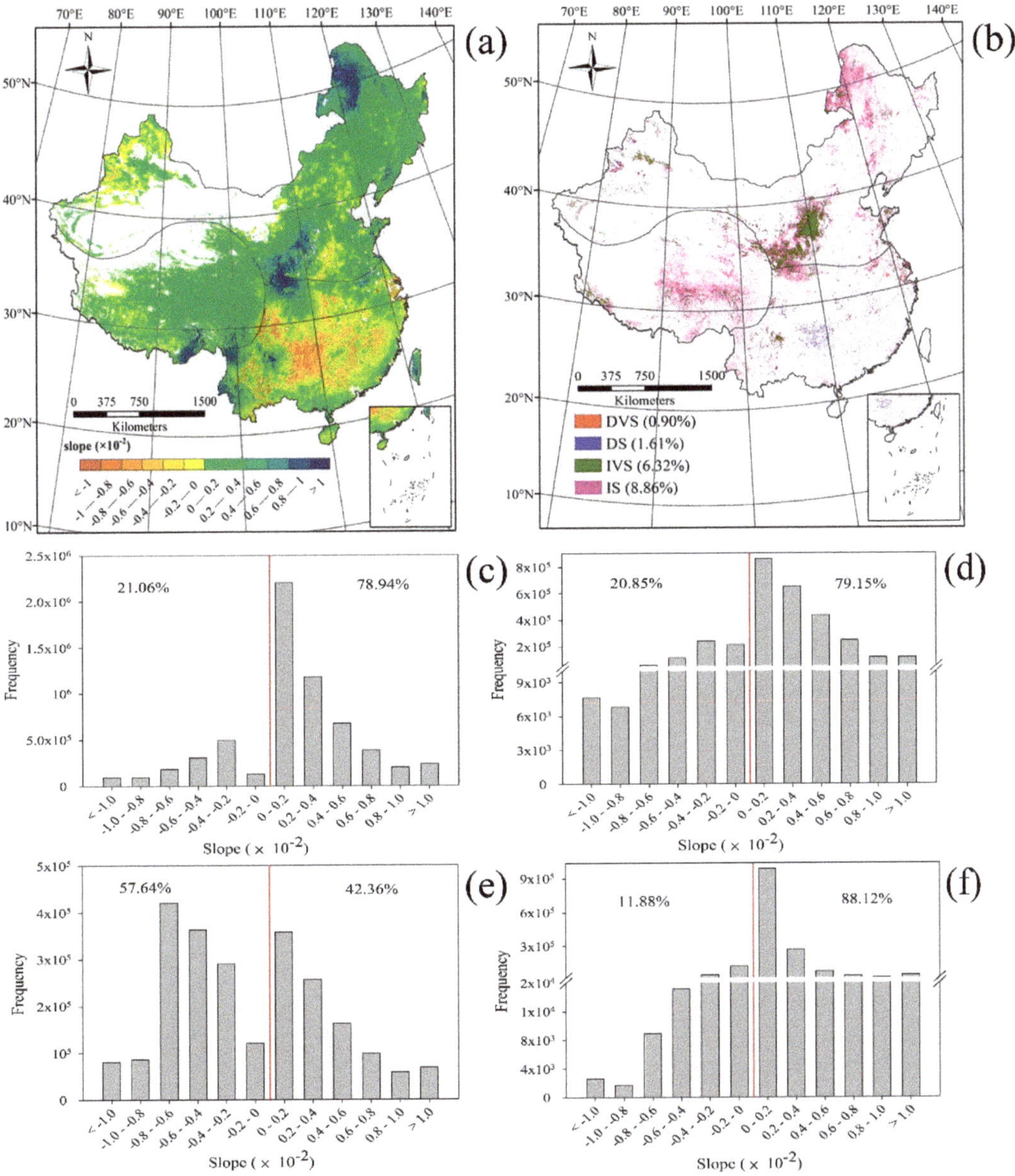

Figure 4. Trends in annual Net Primary Productivity (NPP) within China between 2001 and 2014 (**a**); and pixels with significant ($p < 0.05$) and very significant ($p < 0.01$) trends are shown in (**b**). The count distributions of NPP trends in China (**c**); Northern China (**d**); Southern China (**e**) and the Tibetan Plateau (**f**) are also shown. A positive trend indicated an increase in NPP and vice versa; DS and DVS represent decreased significantly and very significantly, respectively; and IS and IVS represent increased significantly and very significantly, respectively. The red lines in figures (**c**), (**d**), (**e**) and (**f**) represent boundary for negative slop and positive slop.

3.4. Turning Point Year of NPP

3.4.1. Turning Point Year of NPP at National and Regional Scale

To clarify the characteristics of the NPP trends, the averaged trends of the NPP and the turning point year at national and regional levels were calculated over the study period (Figure 5). At the pixel scale, 87.26% of the total pixels showed abrupt changes in inter-annual NPP variation (ΔAIC < −2; Figure S3b), most of which appeared at year 2004 or year 2010 (which may also have been caused by criterion we set up in the calculation), accounting for 29.36% and 31.78%, respectively (Figure S3a). At a national scale, the linear regression model suggested an increasing trend of 0.0007 kg C m^{-2} a^{-1} (p = 0.3118) from 2001 to 2014 (Figure 5a). However, two distinct periods were clearly identified by the piecewise linear regression model: a trend change from a marginal significant increasing trend of 0.0104 kg C m^{-2} a^{-1} (p = 0.0746) before 2004 to a non-significant increasing trend of 0.0012 kg C m^{-2} a^{-1} (p = 0.2775) after 2004 (Figure 5a). The information criterion of the piecewise regression model was less than that of the linear regression model (ΔAIC = −3.1797). At a regional scale, a marginal significant increasing trend of 0.0025 kg C m^{-2} a^{-1} (p = 0.0509) in Northern China was found, and the turning point year occurred in 2010 (ΔAIC < −2). However, the increasing trends were all non-significant before and after 2010, with a rate of 0.0006 and 0.0068 kg C m^{-2} a^{-1}, respectively (Figure 5b). In Southern China, a rate of −0.0007 kg C m^{-2} a^{-1} (p = 0.6559) was found for annual NPP variation, and a turning point year occurred in 2010 (ΔAIC < −2). However, the variation trend was opposite (Figure 5c). Although a significant increasing trend of 0.0016 kg C m^{-2} a^{-1} (p < 0.01) was found in the Tibetan Plateau over the past 14 years (Figure 5d), the increasing trend slowed down after 2006 (ΔAIC < −2).

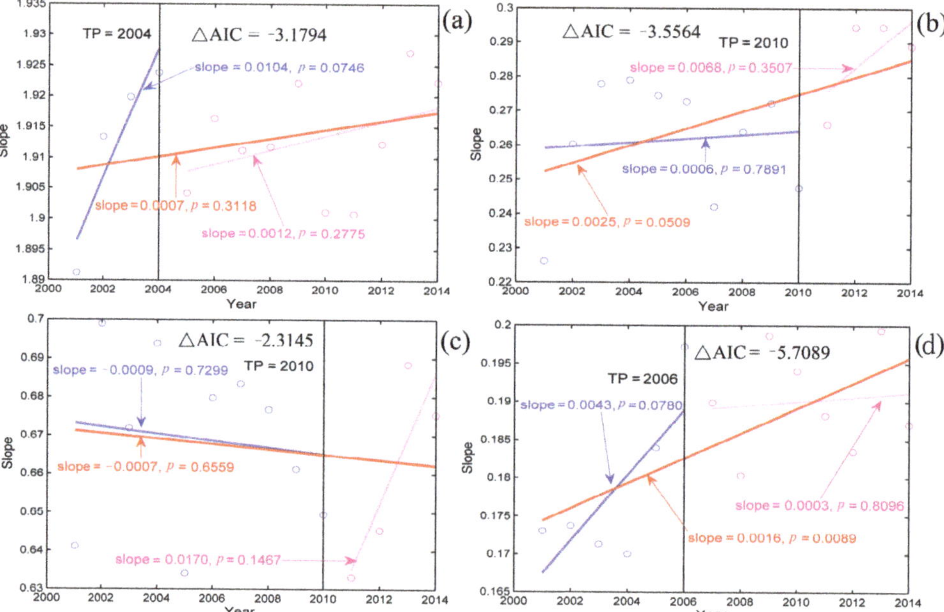

Figure 5. Inter-annual variations of Net Primary Productivity (NPP) in China at national level (**a**) and regional level (Northern China (**b**); Southern China (**c**); and the Tibetan Plateau (**d**)). Trends estimated by the least-squares linear regression are shown. The blue and pink lines indicate the linear fit before and after the turning year, respectively. The red line indicates linear fit during the period 2001–2014. TP represents the turning point year.

3.4.2. Turning Point Year of NPP at Biome Scale

At the biome level, the variation of NPP was diverse (Figure 6). All vegetation types had two clearly distinct periods over the study period (ΔAIC < -2), with the exception of the EBF (ΔAIC = -1.9314) which only showed a non-significant decreasing trend of -0.0001 kg C m^{-2} a^{-1} ($p = 0.9314$) during 2001–2014. Farmland and grassland all increased, with a rate of 0.0011 kg C m^{-2} a^{-1} ($p = 0.3734$) and 0.0016 kg C m^{-2} a^{-1} ($p < 0.01$), respectively. In addition, the increasing trend before 2004 was larger than that after 2004 for both farmland and grassland. An increasing trend of NPP was found for DNF, ENF, and DBF, but the opposite trend was found for shrub, despite a turning point year occurring in 2010 (all ΔAIC < -2 and may also have been caused by the criterion we set up in the calculation) for all above vegetation types. In addition, a decreasing NPP occurred for ENF, DBF and shrub during 2001–2010, while an increasing NPP occurred for DNF during the same time period. After 2010, the trend of these four vegetation types was all positive, despite the existence of different magnitudes.

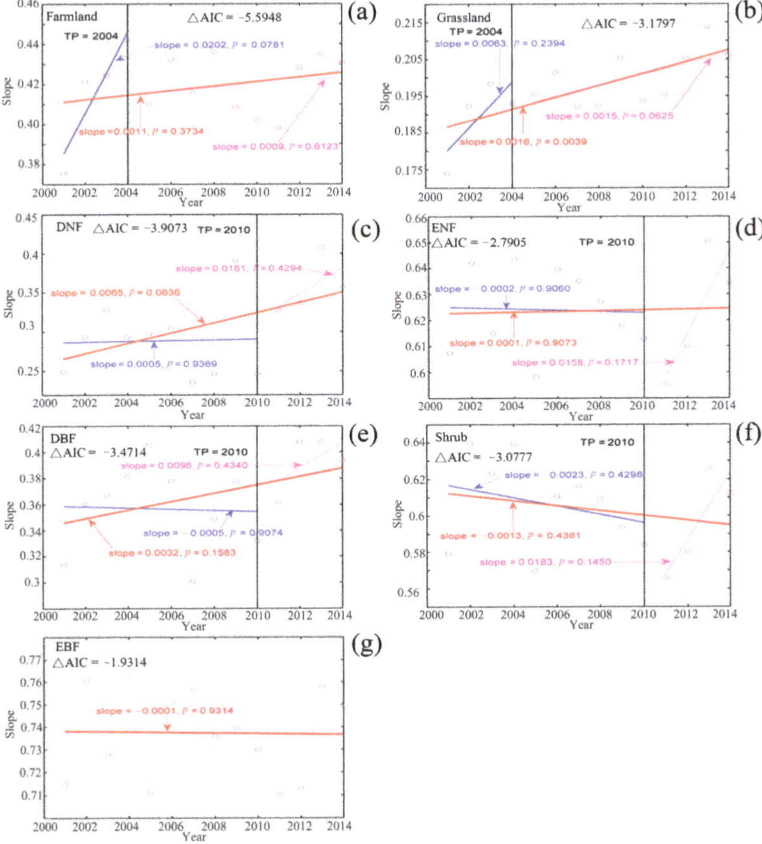

Figure 6. Inter-annual variations of Net Primary Productivity (NPP) in China at the biome level (Farmland (**a**), Grassland (**b**), Deciduous needle-leaf forest (**c**), Evergreen needle-leaf forest (**d**), Deciduous broadleaf forest (**e**), Shrub (**f**), and Evergreen broadleaf forest (**g**)). Trends estimated by the least-squares linear regression are shown. The blue and pink lines indicate the linear fit before and after the turning year, respectively. The red line indicates linear fit during the period 2001–2014. TP represents the turning point year.

3.5. Relationship Relating Climatic Factors, LOS and NPP

3.5.1. Correlation Analysis

Climatic factors were generally positively correlated with NPP on a national scale (Figure 7). Positive partial correlations were observed for 62.66%, 70.79% and 57.60% of the total pixels with temperature, precipitation, and solar radiation, respectively. The positive correlation coefficients for the climatic variables ranged from 0.1–0.7. About 56.23% of the total pixels showed positive partial correlations between NPP and LOS, and the positive partial correlation coefficient was mostly distributed between 0.1 and 0.5.

At a regional level, 62.97% of the total pixels in Northern China displayed a negative relationship between NPP and temperature, while precipitation, solar radiation and LOS all generally showed a positive relationship with the NPP (Table 1). Over 57% of the total pixels in Southern China indicated a positive relationship between the NPP and climatic factors as well as LOS. However, the relationship in the Tibetan Plateau was diverse. Specifically, over 83% of the total pixels for temperature and about 58% of the total pixels for precipitation displayed a positive relationship with the NPP, while in general, an opposite relationship occurred between the NPP and solar radiation. In addition, the relationship between the NPP and LOS in the Tibetan Plateau was ambiguous.

Table 1. Partial correlation between the Net Primary Productivity (NPP) and climatic factors as well as Length of Season (LOS).

Variable	Relationship	Northern China	Southern China	The Tibetan Plateau
Temperature	+	37.03%	67.56%	83.14%
	−	62.97%	32.44%	16.86%
Precipitation	+	84.55%	62.04%	57.83%
	−	15.55%	37.96%	42.17%
Solar radiation	+	66.93%	61.06%	34.75%
	−	33.07%	38.94%	65.25%
LOS	+	57.83%	57.56%	50.73%
	−	42.17%	42.14%	49.27%

+ Represent positive relationship; − Represent negative relationship.

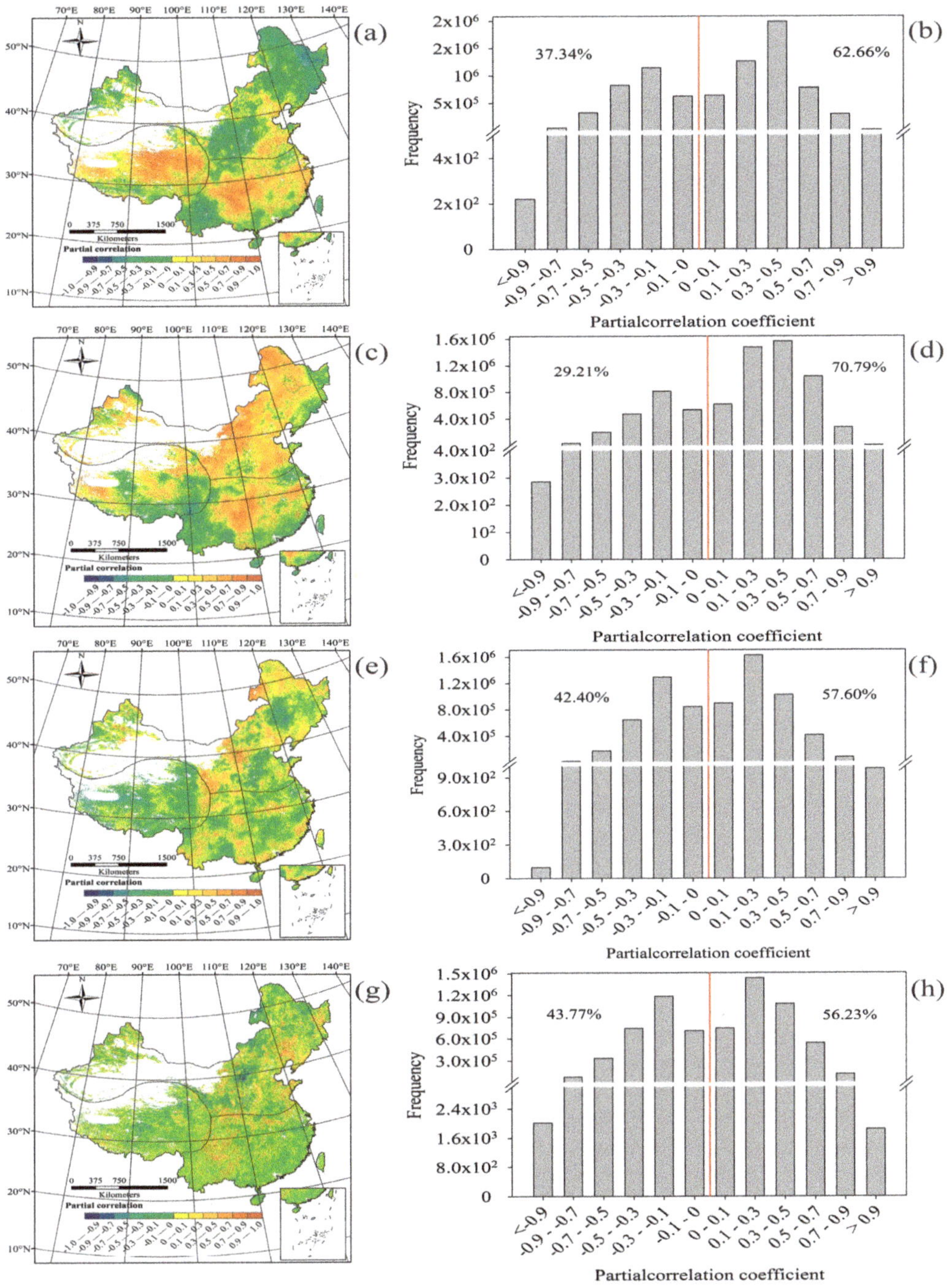

Figure 7. The partial correlation coefficients and corresponding frequency distribution between the NPP and temperature (**a**,**b**), precipitation (**c**,**d**), solar radiation (**e**,**f**) and LOS (**g**,**h**) in China from 2001 to 2014. The red lines in figures (b), (d), (f) and (h) represent boundary for negative relationship and positive relationship.

3.5.2. Structural Equation Models

Our structural equation model (SEM) did not detect a relationship between temperature, precipitation and solar radiation, which indicates that there was a strong relationship among the climatic factors, LOS and NPP. In the SEM (χ^2 = 1.0138, d.f. = 5, p = 0.9614, AIC = 21.0138, BIC = $-$1815), climatic factors explained a total of 23% of the variation in LOS, and temperature and solar radiation had a positive effect on LOS, while precipitation had the opposite effect (Figure 8). Moreover, the effect of climatic factors and LOS on NPP was 0.63, and explained a total of 56% of the variation in NPP. Among the climatic factors, the direct effect of solar radiation on NPP was the largest (0.31), followed by precipitation (0.25), and temperature (0.07).

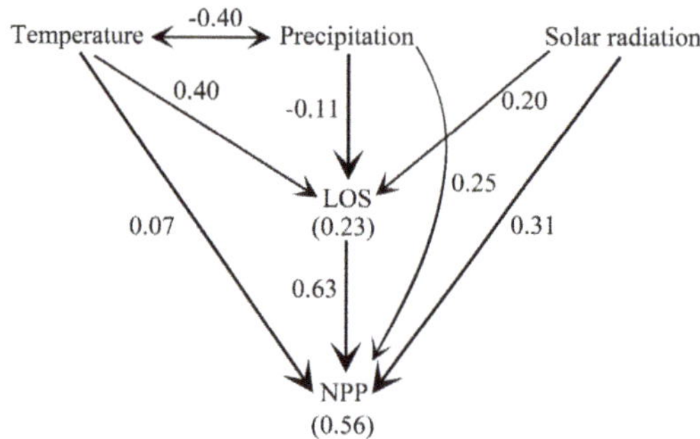

Figure 8. Structural equation model relating climatic factors, Length of Season (LOS) and Net Primary Productivity (NPP) in China. Single headed arrows indicate directional relationships, while double headed arrows indicate covariances. Numbers in brackets are R^2 values.

4. Discussion

4.1. Uncertainties in NPP Estimates

Net primary productivity (NPP), an indicator of the accumulation of atmospheric CO_2 in terrestrial ecosystems, plays a crucial role in global change [54]. The accurate estimation of NPP is a crucial step for reliably quantifying carbon fluxes between the atmosphere and terrestrial ecosystems [31], especially for regions with different topography and climatic conditions. Although deficiency and biases exist in the MODIS derived NPP, the validation illustrated that the model applied by MODIS was superior to the Miami model and Thornthwaite Memorial model, with an increased R^2 value and a decreased RMSE and MAE. The RMSE and MAE between the MODIS derived NPP and observations were only 0.19 kg C m^{-2} and 0.14 kg C m^{-2}, respectively. This indicated that the MODIS derived NPP may be suitable for analysis with relatively low regional biases in China for the period of 2001–2014 to some extent despite coarse global climate data applied in MOD17. Additionally, it illustrated that the parameters input and algorithms may affect the results. For instance, the Miami model is one of the first global empirical models that only utilizes temperature or precipitation in the model [55], while the Thornthwaite Memorial model determines NPP for a particular location as the actual evapotranspiration functions [36]. The interactions among climatic factors and other influential factors such as vegetation types are not included in these two models [11]. Furthermore, there is no mechanism to account for changing vegetation density in the two models [56]. However, the model applied by MODIS incorporates biogeochemical principles in a mechanistic modeling environment and the vegetation feedback to climate conditions through changes in Leaf Area Index and absorbed radiation, and has the advantage of providing spatially continuous estimates with a consistent methodology, which is important for any large-scale studies [35]. It should be noted that NPP is not easy to measure and only 26 field sites were included in this study, which may cause biases due to the larger study area. In addition, heterogeneity, stand density, and pixel resolution may also affect validation results. Therefore, increased observation sites and an extended observation period has become essential for NPP estimation. Additionally, forest inventory data is the only data source that provides large-scale consistent productivity assessments [57]. Therefore, the application of forest inventory data rather than numerous field observed data in this study would be more preferable to validate the suitability of MODIS NPP in China, and more efforts should be made in the future. What is more, temporal analysis of match between models and observations based on a long time period were also a good choice for NPP estimates and comparison.

4.2. Spatiotemporal Variation of NPP

In general, all three NPP estimates showed similar spatial patterns, with mean annual NPP higher in the southeast and gradually decreasing towards the northwest (Figure 3 and Figure S2). This finding was consistent with previous studies [10–12,58]. This may have been due to the climatic gradient (e.g., temperature and precipitation) from the southeast to the northwest in China, which is more favorable for vegetation growth in southern China [11,59]. The lowest NPP value in northwest China and the Tibetan Plateau, which was consistent with References [10,11], may have resulted from low temperature and the absence of precipitation. In addition, vegetation type was also a key factor that affected NPP spatial distribution. In Southern China, the dominant vegetation type are evergreen forests (Figure S1) which have a higher productivity than that of the grassland and desert vegetation (Figure 4) distributed mostly in Northern China and the Tibetan Plateau. Productivity in the southeast Tibetan Plateau was higher than that of other areas in the Tibetan Plateau as demonstrated by our conclusion above.

The total amount of NPP increased at a rate of 0.009 Pg C a^{-1} in China from 2001–2014, which was similar with Reference [10], who found a rate of 0.008 Pg C a^{-1} in China from 1999–2010. The decreases in NPP over Southern China were more than offset by increases in NPP over Northern China and the Tibetan Plateau. The overall increased NPP could help to mitigate climate warming at regional scale and/or global scale to some extent. In terms of trend variation, a continually increasing trend of NPP in China from 2001–2014 was found, despite the occurrence of a turning point year which was similar with the previous finding in Reference [10]. This may be due to the fact that over the past three decades, the central government of China has decided to combat severe environmental degradation, including declining vegetation cover and expanding desertification [29]. To realize this goal, the Three North Shelter Forest System project, the Beijing-Tianjin Sand Source Control Program, and the Grain for Green Project [29] were implemented. Alternatively, as the turning point is related to ENSO events/cycles, it is necessary to validate this conclusion in the future.

At a regional scale, NPP trends in China showed a prominent geographical heterogeneity. The trends before and after the turning point year in Northern China and the Tibetan Plateau were all positive. In contrast, the trend in Southern China was the opposite, which may have been due to ecorestoration, forest/grassland protection, and reforestation in Northern China and the Tibetan Plateau after 2000 [29]. However, in Southern China, freezing low temperatures in early 2008, the severe drought in 2009 [31], and decreased solar radiation [12] may have all contributed to the decreasing NPP. Due to the physiological and/or local environmental conditions, the trend and turning point year among vegetation types were diverse.

4.3. Diverse NPP Correlations with Climate Drivers and LOS

Vegetation NPP is influenced by a variety of factors, and the primary one is climate-related. Temperature, precipitation, and solar radiation are the main climate drivers for vegetation growth [59], but the relationships vary with spatial scale. Nevertheless, limited efforts have been made to investigate the relative roles of different climate variables in the NPP [10]. Generally, temperature has been found to be positively correlated with the NPP at a national level, which is consistent with previous studies [10,60]. This response may in part be related to the increased activity of photosynthetic enzymes [61] and improved capacity for photosynthesis and growth [62]. However, the relationship between temperature and NPP in Northern China was negative. This may have resulted from higher temperatures leading to increased water scarcity caused by accelerated evaporation, which would work against vegetation growth, especially for arid and semi-arid areas in Northern China [49]. Additionally, the positive relationship between precipitation and NPP in Northern China illustrated this conclusion. Precipitation showed a positive correlation with NPP at both the national scale and regional scale, which may have occurred as more precipitation increases soil moisture, satisfying the water requirements for vegetation growth and productivity increase, especially in late spring and summer, which usually have higher temperatures and more evapotranspiration [63].

Little is known about the light-related physiological mechanisms that regulate NPP due to the difficulty in distinguishing the effects of day length (i.e., photoperiod) and light intensity [18], especially when studies are extended to consider their correlations with temperature and precipitation. Therefore, in this study, solar radiation was considered as an integrated surrogate for both day length and sunlight intensity [18] to investigate the effects of solar radiation on the NPP. A positive correlation was observed between the NPP and solar radiation at both the national and regional scales, except for a negative relationship which occurred in the Tibetan Plateau. Generally, increased solar radiation provides sufficient materials and energy for vegetation photosynthesis and solar radiation is typically accompanied by warmer temperatures and sufficient sunlight intensity, both of which can enhance photosynthetic capacity and promote vegetation NPP. The negative correlations between solar radiation and NPP in the Tibetan Plateau may be partly related to larger areas of melting seasonal snow and permafrost soils caused by abundant solar radiation, leading to higher soil moisture content and creating an anaerobic soil environment within the plant root zone, thus limiting vegetation growth [64]. Alternatively, increased solar radiation may enhance surface soil evaporation and limit water availability for herbaceous plants that have shallow root systems.

Vegetation phenology has long been regarded as an important factor that affects vegetation productivity. Consistent with previous studies in References [20,65], our results also indicated a positive relationship between the NPP and LOS at both national and regional levels. This suggests that an extension of LOS is one of the most important factors that affect plant productivity [66].

4.4. Shortcomings and Uncertainties

In this study, we used data from only 99 solar radiation stations, as meteorological stations in the Tibetan Plateau are scarce. The sparse distribution of climate data limited the detail and accuracy of the relationship between climatic factors and plant productivity. Therefore, the results are likely to be subject to some ambiguity.

Vegetation NPP is influenced by a variety of factors. In this study, only three climatic factors and LOS were considered. Although these factors have explained a total of 56% of the variation of NPP (Figure 8), other climate-related factors, such as sunshine duration, soil temperature, precipitation characteristics (e.g., effective precipitation, precipitation intensity), CO_2 concentrations, N enrichment and deposition (e.g., policy and planning changes), should be considered in future studies [67]. Moreover, other phonological metrics (e.g., Start of Season, End of Season), species competition, and disturbances such as anthropogenic activities (irrigation, fertilization [68,69], harvest, land use/land cover change), wildfires, plant diseases, pests, floods, and droughts as well as the time-lag effect of the above–mentioned variables can vary by region [50], and should also be considered in future studies.

The results presented here indicate the complexity of vegetation NPP dynamics and the response or reaction strategy of vegetation to climate change during the study period. However, the trend for NPP before and after turning points were mostly non-significant despite all turning points being significant (except evergreen broadleaf forest) based on ΔAIC. This may be caused by a short time period, or a non-linear variation on NPP during the study period. Therefore, the conclusions mentioned above should be made with more caution, and more effort should be made to explore the NPP trends (e.g., non-linear trend) in the future to clarify the direction of dynamic NPP trends [70]. In addition, it is difficult to estimate future NPP dynamics using mathematical models (e.g., linear regression analysis) due to the nonlinear characteristics of NPP dynamics in a long time series, the uncertainty about future climate change, and the time-lag effect of NPP responses to climate change. To date, the Hurst exponent, which has been widely used in hydrology, climatology, economics, geology, and geochemistry, could be considered to predict the NPP variation. The results presented here (Figure S4) indicate that NPP in China will continue to increase based on the Hurst exponent. Unfortunately, this exponent characterizes trends based only on past and present environmental conditions, without considering future environmental change, especially in developing areas, or giving a time period for dynamic NPP trends. These limitations restrict our ability to predict and anticipate future NPP variation, the carbon cycle and its relationship with and response to climate change and human activity, and hence more effort must be made to predict future NPP variation.

5. Conclusions

In this study, the spatiotemporal patterns of NPP and its correlation with climatic factors as well as vegetation phenology during 2001–2014 in China were investigated, and the main conclusions can be summarized as follows:

1. Validation results showed that NPP derived from MODIS performed well in China's ecosystem compared with the other two models, with an increased R^2 value and a decreased RMSE and MAE. However, this conclusion should be made with caution due to few observed sites.
2. During the entire study period, annual NPP showed an increasing trend at the national scale. However, the increasing trends in NPP were not linear, with a slower rate after 2004. At a regional level, annual NPP displayed an increasing trend, despite the occurrence of a turning point year in Northern China and the Tibetan Plateau. However, the NPP in Southern China decreased during the whole study period, which can mainly be explained by climate change as well as fierce anthropogenic activities.
3. Generally, climatic factors and LOS were positively correlated with NPP, and the direct effect of solar radiation on NPP was the largest compared with temperature and precipitation. Due to local climate conditions, the relationship between NPP and climatic factors were diverse at a regional scale.

Supplementary Materials: The following are available online at www.mdpi.com/1999-4907/8/10/361/s1, Table S1: Sites information of the field Net primary productivity (NPP) data used in this study; Figure S1: Vegetation types (DNF, deciduous needle-leaf forest; ENF, evergreen needle-leaf forest; EBF, evergreen broadleaf forest; DBF, deciduous broadleaf forest; and GM, grassland and meadow) as well as three sub-region divisions (**a**), and provinces distribution (**b**) in China; Figure S2: Distribution of meteorological stations across China, red points represent stations for temperature and precipitation, and black triangles represent stations for solar radiation; Figure S3: Spatial distribution of Turing Point (TP) years (**a**) and corresponding ΔAIC values (**b**) of the NPP in China for the period 2001–2014; Figure S4: Spatial distribution of the Hurst exponent in China from 2001 to 2014.

Acknowledgments: This study is supported by the National Natural Science Foundation of China (Grant No. 41431178), the Natural Science Foundation of Guangdong Province, China (Grant No. 2016A030311016), the Innovation Project of Guangdong Province Water Resources Department (Grant No. 2015-02) and the Central Fund Supporting Nonprofit Scientific Institutes for Basic Research and Development (PM-zx021-201407-007).

Author Contributions: Xinchang Zhang and Xijun Yu outlined the research topic, assisted with manuscript writing, and coordinated the revision activities. Jian Yang and Zhaohui Luo performed data collection, data analysis, the interpretation of results, manuscript writing, and coordinated the revision activities.

Conflicts of Interest: The authors declare no conflict of interest.

References

1. Haberl, H.; Erb, K.H.; Krausmann, F.; Gaube, V.; Bondeau, A.; Plutzar, C.; Gingrich, S.; Lucht, W.; Fischer-Kowalski, M. Quantifying and mapping the human appropriation of net primary production in earth's terrestrial ecosystems. *Proc. Natl. Acad. Sci. USA* **2007**, *104*, 12942–12947. [CrossRef] [PubMed]

2. Allen, C.D.; Macalady, A.K.; Chenchouni, H.; Bachelet, D.; McDowell, N.; Vennetier, M.; Kitzberger, T.; Rigling, A.; Breshears, D.D.; Hogg, E.H.; et al. A global overview of drought and heat-induced tree mortality reveals emerging climate change risks for forests. *For. Ecol. Manag.* **2010**, *259*, 660–684. [CrossRef]

3. Luo, Z.; Tian, D.; Ning, C.; Yan, W.; Xiang, W.; Peng, C. Roles of koelreuteria bipinnata as a suitable accumulator tree species in remediating Mn, Zn, Pb, and Cd pollution on Mn mining wastelands in southern china. *Environ. Earth Sci.* **2015**, *74*, 4549–4559. [CrossRef]

4. Pimm, S.L.; Raven, P. Biodiversity. Extinction by numbers. *Nature* **2000**, *403*, 843–845. [CrossRef] [PubMed]

5. Sala, O.E.; Chapin, F.S.; Armesto, J.J.; Berlow, E.; Bloomfield, J.; Dirzo, R.; Huber-Sanwald, E.; Huenneke, L.F.; Jackson, R.B.; Kinzig, A. Global biodiversity scenarios for the year 2100. *Science* **2000**, *287*, 1770–1774. [CrossRef] [PubMed]

6. Houghton, R.A.; Hackler, J.L.; Lawrence, K.T. The U.S. Carbon budget: Contributions from land-use change. *Science* **1999**, *285*, 574–578. [CrossRef] [PubMed]

7. Twine, T.E.; Kucharik, C.J. Climate impacts on net primary productivity trends in natural and managed ecosystems of the central and eastern united states. *Agric. For. Meteorol.* **2009**, *149*, 2143–2161. [CrossRef]

8. Burnham, K.P.; Anderson, D.R. *Model Selection and Multimodel Inference: A Practical Information-Theoretic Approach*; Springer Science & Business Media: Berlin, Germany, 2003.

9. Sisay, K.; Thurnher, C.; Belay, B.; Lindner, G.; Hasenauer, H. Volume and carbon estimates for the forest area of the amhara region in northwestern ethiopia. *Forests* **2017**, *8*, 122. [CrossRef]

10. Liang, W.; Yang, Y.; Fan, D.; Guan, H.; Zhang, T.; Long, D.; Zhou, Y.; Bai, D. Analysis of spatial and temporal patterns of net primary production and their climate controls in china from 1982 to 2010. *Agric. For. Meteorol.* **2015**, *204*, 22–36. [CrossRef]

11. Zhang, M.; Lal, R.; Zhao, Y.; Jiang, W.; Chen, Q. Estimating net primary production of natural grassland and its spatio-temporal distribution in china. *Sci. Total Environ.* **2016**, *553*, 184–195. [CrossRef] [PubMed]

12. Wang, J.; Dong, J.; Yi, Y.; Lu, G.; Oyler, J.; Smith, W.; Zhao, M.; Liu, J.; Running, S. Decreasing net primary production due to drought and slight decreases in solar radiation in China from 2000 to 2012. *J. Geophys. Res. Biogeosci.* **2017**, *122*, 261–278. [CrossRef]

13. Pan, Y.; Wu, J.; Xu, Z. Analysis of the tradeoffs between provisioning and regulating services from the perspective of varied share of net primary production in an alpine grassland ecosystem. *Ecol. Complex.* **2014**, *17*, 79–86. [CrossRef]

14. Brouwers, N.C.; Coops, N.C. Decreasing net primary production in forest and shrub vegetation across southwest australia. *Ecol. Indic.* **2016**, *66*, 10–19. [CrossRef]

15. Zhao, M.; Heinsch, F.A.; Nemani, R.R.; Running, S.W. Improvements of the modis terrestrial gross and net primary production global data set. *Remote Sens. Environ.* **2005**, *95*, 164–176. [CrossRef]

16. The Modis Land Cover and Land Cover Dynamics Products. Available online: https://xue.glgoo.net/scholar?cluster=16597624334725163726&hl=zh-CN&as_sdt=2005&sciodt=0,5 (accessed on 22 June 2017).

17. Gu, F.; Zhang, Y.; Huang, M.; Tao, B.; Guo, R.; Yan, C. Effects of climate warming on net primary productivity in China during 1961–2010. *Ecol. Evol.* **2017**. [CrossRef] [PubMed]

18. Calle, Z.; Schlumpberger, B.O.; Piedrahita, L.; Leftin, A.; Hammer, S.A.; Tye, A.; Borchert, R. Seasonal variation in daily insolation induces synchronous bud break and flowering in the tropics. *Trees* **2010**, *24*, 865–877. [CrossRef]

19. Luo, Z.; Yu, S. Spatiotemporal variability of land surface phenology in china from 2001–2014. *Remote Sens.* **2017**, *9*, 65. [CrossRef]

20. Wu, C.; Hou, X.; Peng, D.; Gonsamo, A.; Xu, S. Land surface phenology of china's temperate ecosystems over 1999–2013: Spatial-temporal patterns, interaction effects, covariation with climate and implications for productivity. *Agric. For. Meteorol.* **2016**, *216*, 177–187. [CrossRef]

21. Peng, J.; Liu, Z.; Liu, Y.; Wu, J.; Han, Y. Trend analysis of vegetation dynamics in qinghai–tibet plateau using hurst exponent. *Ecol. Indic.* **2012**, *14*, 28–39. [CrossRef]

22. Zhang, Y.; Gao, J.; Liu, L.; Wang, Z.; Ding, M.; Yang, X. Ndvi-based vegetation changes and their responses to climate change from 1982 to 2011: A case study in the koshi river basin in the middle himalayas. *Glob. Planet. Chang.* **2013**, *108*, 139–148. [CrossRef]

23. Wen, Z.; Wu, S.; Chen, J.; Lü, M. Ndvi indicated long-term interannual changes in vegetation activities and their responses to climatic and anthropogenic factors in the three gorges reservoir region, China. *Sci. Total Environ.* **2017**, *574*, 947–959. [CrossRef] [PubMed]

24. Piao, S.; Ciais, P.; Lomas, M.; Beer, C.; Liu, H.; Fang, J.; Friedlingstein, P.; Huang, Y.; Muraoka, H.; Son, Y. Contribution of climate change and rising CO_2 to terrestrial carbon balance in east asia: A multi-model analysis. *Glob. Planet. Chang.* **2011**, *75*, 133–142. [CrossRef]

25. Mu, Q.; Zhao, M.; Running, S.W.; Liu, M.; Tian, H. Contribution of increasing CO_2 and climate change to the carbon cycle in China's ecosystems. *J. Geophys. Res. Biogeosci.* **2008**. [CrossRef]

26. Xu, W.C.; Wang, W.; Ma, J.S.; Yang, X. The characteristics, causes of formation and climatic impact of the 1997–1998 El Niño event. *Donghai Mar. Sci.* **2004**, *22*, 1–8.

27. Lu, E.; Luo, Y.; Zhang, R.; Wu, Q.; Liu, L. Regional atmospheric anomalies responsible for the 2009–2010 severe drought in china. *J. Geophys. Res. Atmos.* **2011**. [CrossRef]

28. Liu, J.; Zhang, Q.; Hu, Y. Regional differences of China's urban expansion from late 20th to early 21st century based on remote sensing information. *Chin. Geogr. Sci.* **2012**, *22*, 1–14. [CrossRef]

29. Duan, H.; Yan, C.; Tsunekawa, A.; Song, X.; Li, S.; Xie, J. Assessing vegetation dynamics in the three-north shelter forest region of china using avhrr ndvi data. *Environ. Earth Sci.* **2011**, *64*, 1011–1020. [CrossRef]

30. Sasai, T.; Saigusa, N.; Nasahara, K.N.; Ito, A.; Hashimoto, H.; Nemani, R.; Hirata, R.; Ichii, K.; Takagi, K.; Saitoh, T.M. Satellite-driven estimation of terrestrial carbon flux over far east asia with 1-km grid resolution. *Remote Sens. Environ.* **2011**, *115*, 1758–1771. [CrossRef]

31. Liu, Y.; Ju, W.; He, H.; Wang, S.; Sun, R.; Zhang, Y. Changes of net primary productivity in china during recent 11 years detected using an ecological model driven by modis data. *Front. Earth Sci.* **2013**, *7*, 112–127. [CrossRef]

32. Liu, F.H.; Chen, X.; Chen, X.W.; Song, S. Relationship between temperature change and in climate boundary and summer precipitation over the huaihe river basin. *Clim. Environ. Res.* **2010**, *15*, 169–178.

33. Wang, H.; Dai, J.; Zheng, J.; Ge, Q. Temperature sensitivity of plant phenology in temperate and subtropical regions of china from 1850 to 2009. *Int. J. Climatol.* **2015**, *35*, 913–922. [CrossRef]

34. Shen, M.; Tang, Y.; Chen, J.; Zhu, X.; Zheng, Y. Influences of temperature and precipitation before the growing season on spring phenology in grasslands of the central and eastern qinghai-tibetan plateau. *Agric. For. Meteorol.* **2011**, *151*, 1711–1722. [CrossRef]

35. Zhao, M.; Running, S.W. Drought-induced reduction in global terrestrial net primary production from 2000 through 2009. *Science* **2010**, *329*, 940–943. [CrossRef] [PubMed]

36. Lieth, H. Modeling the primary productivity of the world. In *Primary Productivity of the Biosphere*; Lieth, H., Whittaker, R.H., Eds.; Springer: Berlin/Heidelberg, Germany, 1975; pp. 237–263.

37. ChinaFLUX Sites. Available online: http://www.chinaflux.org/ (accessed on 10 September 2017).

38. Baldocchi, D.; Falge, E.; Gu, L.; Olson, R.; Hollinger, D.; Running, S.; Anthoni, P.; Bernhofer, C.; Davis, K.; Evans, R.; et al. Fluxnet: A new tool to study the temporal and spatial variability of ecosystem–scale carbon dioxide, water vapor, and energy flux densities. *Bull. Am. Meteorol. Soc.* **2001**, *82*, 2415–2434. [CrossRef]

39. China Meteorological Data Sharing Service System. Available online: http://cdc.cma.gov.cn/ (accessed on 6 June 2016).

40. Xu, L.; Chen, X. Regional unified model-based leaf unfolding prediction from 1960 to 2009 across northern china. *Glob. Chang. Biol.* **2013**, *19*, 1275–1284. [CrossRef] [PubMed]

41. Peng, S.; Chen, A.; Xu, L.; Cao, C.; Fang, J.; Myneni, R.B.; Pinzon, J.E.; Tucker, C.J.; Piao, S. Recent change of vegetation growth trend in China. *Environ. Res. Lett.* **2011**. [CrossRef]

42. Zhu, W.; Tian, H.; Xu, X.; Pan, Y.; Chen, G.; Lin, W. Extension of the growing season due to delayed autumn over mid and high latitudes in north america during 1982–2006. *Glob. Ecol. Biogeogr.* **2012**, *21*, 260–271. [CrossRef]

43. Jönsson, P.; Eklundh, L. *Timesat 3.0 Software Manual*; Department of Earth and Ecosystem Sciences, Lund University and Center for Technology Studies, Malmö University: Malmö, Sweden, 2010.

44. Sun, L.; Wei, J.; Duan, D.H.; Guo, Y.M.; Yang, D.X.; Jia, C.; Mi, X.T. Impact of land-use and land-cover change on urban air quality in representative cities of china. *J. Atmos. Sol.-Terr. Phys.* **2016**, *142*, 43–54. [CrossRef]

45. Wei, Y.-X.; Wang, L.-W. Progress in research on land cover products of modis. *Spectrosc. Spectr. Anal.* **2010**, *30*, 1848–1852.

46. Neeti, N.; Eastman, J.R. A contextual mann-kendall approach for the assessment of trend significance in image time series. *Trans. GIS* **2011**, *15*, 599–611. [CrossRef]

47. Wang, C.; Guo, H.; Zhang, L.; Liu, S.; Qiu, Y.; Sun, Z. Assessing phenological change and climatic control of alpine grasslands in the tibetan plateau with modis time series. *Int. J. Biometeorol.* **2015**, *59*, 11–23. [CrossRef] [PubMed]

48. Zhou, J.; Cai, W.; Qin, Y.; Lai, L.; Guan, T.; Zhang, X.; Jiang, L.; Du, H.; Yang, D.; Cong, Z. Alpine vegetation phenology dynamic over 16 years and its covariation with climate in a semi-arid region of china. *Sci. Total Environ.* **2016**, *572*, 119–128. [CrossRef] [PubMed]

49. Xu, H.-J.; Wang, X.-P.; Yang, T.-B. Trend shifts in satellite-derived vegetation growth in central eurasia, 1982–2013. *Sci. Total Environ.* **2017**, *579*, 1658–1674. [CrossRef] [PubMed]

50. Xu, G.; Zhang, H.; Chen, B.; Zhang, H.; Innes, J.L.; Wang, G.; Yan, J.; Zheng, Y.; Zhu, Z.; Myneni, R.B. Changes in vegetation growth dynamics and relations with climate over China's landmass from 1982 to 2011. *Remote Sens.* **2014**, *6*, 3263–3283. [CrossRef]

51. Akaike, H. A new look at the statistical model identification. *IEEE Trans. Autom. Control* **1974**, *19*, 716–723. [CrossRef]

52. Shen, Y.; Yu, S.; Lian, J.; Shen, H.; Cao, H.; Lu, H.; Ye, W. Tree aboveground carbon storage correlates with environmental gradients and functional diversity in a tropical forest. *Sci. Rep.* **2016**, *6*, 25304. [CrossRef] [PubMed]

53. Fox, J. Teacher's corner: Structural equation modeling with the sem package in R. *Struct. Equ. Model. Multidiscip. J.* **2006**, *13*, 465–486. [CrossRef]

54. Roxburgh, S.H.; Berry, S.L.; Buckley, T.N.; Barnes, B.; Roderick, M.L. What is NPP? Inconsistent accounting of respiratory fluxes in the definition of net primary production. *Funct. Ecol.* **2005**, *19*, 378–382. [CrossRef]

55. Zaks, D.P.M.; Ramankutty, N.; Barford, C.C.; Foley, J.A. From miami to madison: Investigating the relationship between climate and terrestrial net primary production. *Glob. Biogeochem. Cycles* **2007**. [CrossRef]

56. Adams, B.; White, A.; Lenton, T.M. An analysis of some diverse approaches to modelling terrestrial net primary productivity. *Ecol. Model.* **2004**, *177*, 353–391. [CrossRef]

57. Tomppo, E.; Gschwantner, T.; Lawrence, M.; McRoberts, R. *National Forest Inventories: Pathways for Common Reporting*; Springer: Berlin, Germany, 2010.

58. Piao, S.; Fang, J.; Zhou, L.; Zhu, B.; Tan, K.; Tao, S. Changes in vegetation net primary productivity from 1982 to 1999 in china. *Glob. Biogeochem. Cycles* **2005**, *19*, 1–16. [CrossRef]

59. Richardson, A.D.; Keenan, T.F.; Migliavacca, M.; Ryu, Y.; Sonnentag, O.; Toomey, M. Climate change, phenology, and phenological control of vegetation feedbacks to the climate system. *Agric. For. Meteorol.* **2013**, *169*, 156–173. [CrossRef]

60. Yu, G.-R.; Zhu, X.-J.; Fu, Y.-L.; He, H.-L.; Wang, Q.-F.; Wen, X.-F.; Li, X.-R.; Zhang, L.-M.; Zhang, L.; Su, W.; et al. Spatial patterns and climate drivers of carbon fluxes in terrestrial ecosystems of China. *Glob. Chang. Biol.* **2013**, *19*, 798–810. [CrossRef] [PubMed]

61. Shi, C.; Sun, G.; Zhang, H.; Xiao, B.; Ze, B.; Zhang, N.; Wu, N. Effects of warming on chlorophyll degradation and carbohydrate accumulation of alpine herbaceous species during plant senescence on the tibetan plateau. *PLoS ONE* **2014**, *9*, e107874. [CrossRef] [PubMed]

62. Liu, Q.; Fu, Y.H.; Zhu, Z.; Liu, Y.; Liu, Z.; Huang, M.; Janssens, I.A.; Piao, S. Delayed autumn phenology in the northern hemisphere is related to change in both climate and spring phenology. *Glob. Chang. Biol.* **2016**, *22*, 3702–3711. [CrossRef] [PubMed]

63. Ibrahim, Y.; Balzter, H.; Kaduk, J.; Tucker, C. Land degradation assessment using residual trend analysis of gimms ndvi3g, soil moisture and rainfall in sub-saharan west africa from 1982 to 2012. *Remote Sens.* **2015**, *7*, 5471–5494. [CrossRef]

64. Yang, Y.; Guan, H.; Shen, M.; Liang, W.; Jiang, L. Changes in autumn vegetation dormancy onset date and the climate controls across temperate ecosystems in china from 1982 to 2010. *Glob. Chang. Biol.* **2015**, *21*, 652–665. [CrossRef] [PubMed]

65. Dragoni, D.; Schmid, H.P.; Wayson, C.A.; Potter, H.; Grimmond, C.S.B.; Randolph, J.C. Evidence of increased net ecosystem productivity associated with a longer vegetated season in a deciduous forest in south-central indiana, USA. *Glob. Chang. Biol.* **2011**, *17*, 886–897. [CrossRef]

66. Piao, S.; Friedlingstein, P.; Ciais, P.; Viovy, N.; Demarty, J. Growing season extension and its impact on terrestrial carbon cycle in the northern hemisphere over the past 2 decades. *Glob. Biogeochem. Cycles* **2007**. [CrossRef]

67. Suepa, T.; Qi, J.; Lawawirojwong, S.; Messina, J.P. Understanding spatio-temporal variation of vegetation phenology and rainfall seasonality in the monsoon southeast asia. *Environ. Res.* **2016**, *147*, 621–629. [CrossRef] [PubMed]
68. He, L.; Chen, J.M.; Liu, J.; Bélair, S.; Luo, X. Assessment of smap soil moisture for global simulation of gross primary production. *J. Geophys. Res. Biogeosci.* **2017**, *122*, 1549–1563. [CrossRef]
69. Liu, X.; Zhang, Y.; Han, W.; Tang, A.; Shen, J.; Cui, Z.; Vitousek, P.; Erisman, J.W.; Goulding, K.; Christie, P.; et al. Enhanced nitrogen deposition over china. *Nature* **2013**, *494*, 459–462. [CrossRef] [PubMed]
70. Li, S.C.; Zhao, Z.Q.; Gao, Y.; Wang, Y.L. Determining the predictability and the spatial pattern of urban vegetation using recurrence quantification analysis: A case study of shenzhen city. *Geogr. Res.* **2008**, *27*, 1243–1252.

 *forests*

Article

Plant-Soil Properties Associated with Nitrogen Mineralization: Effect of Conversion of Natural Secondary Forests to Larch Plantations in a Headwater Catchment in Northeast China

Qiong Wang [1,2], **Fayun Li** [1,2,*], **Xiangmin Rong** [1] **and Zhiping Fan** [2]

1   College of Resources and Environment, Hunan Agricultural University, No. 1 Nongda Road, Changsha 410128, China; wangqiong0407@163.com (Q.W.); rongxm2005@126.com (X.R.)
2   Institute of Eco-environmental Sciences, Liaoning Shihua University, No.1 Dandong West Road, Fushun 113001, China; zhiping_fan@hotmail.com
*   Correspondence: lifayun15@hotmail.com; Tel.: +86-24-5686-3960

Received: 8 May 2018; Accepted: 25 June 2018; Published: 28 June 2018

Abstract: To understand the relative importance of plant community composition and plant-induced soil properties on N transformations, the soil N mineralization, ammonification and nitrification characteristics of natural secondary forests (*Quercus mongolica-Juglans mandshurica* forest: QJF, and *Quercus mongolica-Populus davidiana* forest: QPF) and the adjacent larch plantations (*Larix kaempferi* forest: LF1 and LF2) were studied during the growing season. All of the forest types showed seasonal dynamics of N mineralization rates. The total cumulative N mineralization was significantly higher in QPF (73.51 kg hm^{-2}) than in LF1 (65.64 kg hm^{-2}) and LF2 (67.51 kg hm^{-2}) ($p < 0.05$). The total cumulative nitrification from May to November was significantly higher in QJF (65.16 kg hm^{-2}) and QPF (64.87 kg hm^{-2}) than in LF1 (52.62 kg hm^{-2}) and FL2 (54.17 kg hm^{-2}) ($p < 0.05$). Based on the variation partitioning, independent soil properties were the primary determinants of the N transformations (13.5%). Independent climate conditions explained 5.6% of the variations, while plant variations explained 3.2% of the variations in N transformations. We concluded that different forest types with various plant community compositions have different influences on the litterfall quantity and quality and the nutrient availability, and these differences interact with seasonal climate conditions that in turn drive the differences in N mineralization.

Keywords: soil nitrogen mineralization; plant-soil interactions; resin core method; forest conversion; headwater catchment

1. Introduction

Nitrogen (N) is an essential element for the growth of organisms and the productivity of forest ecosystems [1]. In headwater catchments, soil N mineralization of organic matter plays an important role in determining soil N availability, primary productivity and N losses from soil to stream, thus contributing to ground water contamination and the pollution of the water environment [1–4]. Factors affecting the temporal and spatial patterns of soil N dynamics have been well documented [5–7]. Numerous previous studies and practices have shown that seasonal changes in N mineralization result in patterns with the highest mineralization rates in the summer and the lowest rates in the winter, which appears to follow seasonal patterns of temperature and precipitation [8–11]. Soil temperature, moisture, and precipitation patterns are important drivers of soil N transformations, and each of these seasonal climate conditions may have different impacts on various forest types [12]. Forest types with varying plant communities may have different influences on the N cycle due to differences in the physiology, morphology, nutrient requirements, and life histories of various plant species [13–15].

There are several approaches of studying the N mineralization for different forest types, but they generally do not consider plant community compositions and species diversity. Conversion of natural forests to plantations often leads to considerable losses of plant species and consequently a reduction in the diversity of litter species compositions and the amount of litter production, which affects soil nutrient availability and N transformation [16]. Larch (*Larix kaempferi*) plantations are the most widespread forests in northeastern China, but their ecological impacts receive little attention. Therefore, understanding the mechanisms underlying the effects of natural forests converted into plantations on N mineralization of organic matter in headwater catchments is useful for forest management and structure regulation, and can thus help minimize N exports to aquatic ecosystems.

There are three dominant processes between plant communities and soil properties that could explain the mechanisms underlying the effects of natural forests converted into plantations on N mineralization of organic matter. First, changes in plant community composition could influence soil N mineralization via affecting soil nutrient availability, e.g., total nitrogen (TN), soil organic carbon (SOC), C:N ratio, and dissolved organic carbon (DOC), since tree species exhibit differences in the quality of plant material and chemical compounds which significantly affect organic matter input and decomposition. Grime (1998) found that a community dominated by plants with high nitrogen concentrations would likely have positive effects on N mineralization rates [17]. Second, productivity could also influence N mineralization because approximately 50–60% of plant-assimilated N in deciduous forest is annually returned to the soil via litterfall [18]. Denton (1999) and Mikola (2000) found that greater inputs of plant material could increase N mineralization rates because soil microbial biomass and activity have been shown to respond to increased nitrogen and carbon resources [19,20]. Third, plant diversity could also affect N mineralization rates because a more diverse array of plant material entering the soil through leaf litter, fine root production and root turnover could affect N mineralization rates by providing a consistent long-term supply of organic nitrogen as the qualities of plant material decomposed at various rates [21]. These three attributes of the interactions between plant communities and soil processes may simultaneously affect N mineralization. Temperature and precipitation changes are likely to influence N mineralization by altering factors like those discussed above. For example, an increase in temperature can enhance microbial activities and increase the rates of litter decomposition, which, as a result, can change the N mineralization rate. Different plant community compositions have different substrate inputs, soil chemistry and microbial activity, and such differences may contribute unequally to the soil N mineralization; therefore, the N mineralization processes in different forest types are likely to respond differently to seasonal changes in temperature and precipitation.

To understand the relative importance of plant community composition and plant-induced soil property effects of forest conversion on N transformation patterns, organic N mineralization, nitrification and ammonification were investigated in natural secondary forests and the adjacent larch plantations in the headwater catchment of the Taizi River in China. This study aimed to (1) investigate and compare seasonal N mineralization rates under field conditions in natural secondary forests and the adjacent larch plantations; and (2) assess the extent to which N mineralization rates could be explained by the plant-soil properties that are associated with plant community compositions and seasonal climate conditions in temperature and precipitation.

2. Materials and Methods

2.1. Study Area

The Laotudingzi National Nature Reserve (124°41′13″–125°05′15″ E; 41°11′11″–41°21′34″ N) is situated in the headwater catchment of the Taizi River in Liaoning Province, China. The area has a temperate monsoon climate, with mean annual temperature of 6.2 °C and a mean annual rainfall of 778 mm, of which 60–65% falls between June and August. During the study period, the temperature and precipitation largely followed this long-term seasonal pattern. Due to the cold weather during the long winter in Northeast China, soil freezing occurred from November to early April. The average growing season is approximately 215 frost-free days. The air temperature and precipitation from January to December 2014 was measured at a weather station close to the experimental site (Figure 1).

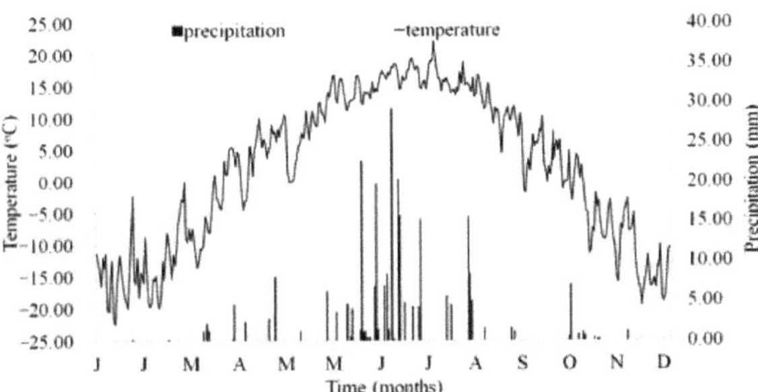

Figure 1. Seasonal dynamics of air temperature and precipitation during 1 year from January to December 2014 in Laotudingzi National Nature Reserve.

The study area had been primarily covered by broadleaf Korean pine forests until the 1930s and thereafter subjected to unregulated timber removal for decades. Massive controlled burns where used in the early 1950s for clearing out the original forest. Since then, the study site has been progressively covered by a naturally regenerated secondary forest. The natural secondary forests consisted of *Quercus mongolica*, *Juglans mandshurica*, *Populus davidiana*, *Acer mono Maxim*, *Phellodendron amurense*, *Fraxinus mandshurica*, *Pinus koraiensis*, *Betula platyphylla* and *Tilia amurensis*. At the beginning of the forest succession, some patches of the natural secondary forests were cleared and replaced by 3-year-old larch (*Larix kaempferi*) seedlings. The larch plantations contain *Larix kaempferi*, *Phellodendron amurense*, *Quercus mongolica*, *Juglans mandshurica* and *Fraxinus mandshurica*.

In this study, two larch plantations sites (LF1 and LF2) and two natural secondary forests (QJF and QPF) were selected. Three independently fixed 20 × 20 m plots were randomly selected at each site. The soil is a typical brown forest soil (classified as Udalfs according to the second edition of USDA soil taxonomy) and with depth 20 to 40 cm. Detailed data on the stands, plots and samples are given in Table 1.

Table 1. Main characteristics of LF1, LF2, QJF and QPF stands.

Item	LF1	LF2	QJF	QPF
Representative plants	*Larix kaempferi*	*Larix kaempferi*	*Quercus mongolica* *Juglans mandshurica*	*Quercus mongolica* *Populus davidiana*
Slope (°)	18	18	16	18
Elevation (m)	656	684	672	663
Forest age (a)	43	43	43	46
No. of tree species	3	8	12	10
Tree density (stems/hm^2)	150	410	383	312
Canopy density	0.6	0.8	0.9	0.9
Diameter at breast height (cm)	24.57 ± 5.98	24.82 ± 3.74	24.53 ± 9.73	25.37 ± 8.00
Tree height (m)	20.16 ± 2.44	20.22 ± 1.53	19.00 ± 3.56	19.88 ± 3.90

2.2. Variables Assessed

2.2.1. Vegetation Survey

All individual trees ≥1 cm in diameter at breast height (DBH) were tallied and recorded by species name, tree height, DBH and canopy density at each plot. Within each plot three sub-plots of 1 × 1 m were laid for herbs, and the number of species, number of individuals per species and coverage in the three sub-plots were recorded. In each sub-plot, forest floor litter, including leaf litter, senesced branches, and bark were collected. We divided the litter layer into two sub-layers: the L layer (undecomposed litter, consisting of litter with clear recognizable structure lying loosely on the forest floor surface) and the F/H layer (mixture of partly decomposed litters where plant remains are partially decomposed by biological activity but with plant morphology still recognizable and humus without recognizable plant structures and a fine granular morphology) [22]. After litter materials were collected, the above and below ground parts of all herbs were harvested through destructive sampling from these sub-plots. To account for the annual litterfall, three samples were randomly collected in each plot every month using a 0.5 m^2 litterfall traps from May to November. Then, all samples were weighed and oven-dried at 65 °C to measure the dry mass and evaporated water content.

The Shannon Weaver Diversity Index was used to measure species diversity [23], and Margalef's Index was used to estimate species richness [24].

Shannon Weaver Diversity Index (H'):

$$\exp(H') = \sum_{i=1}^{R} p_i \ln p_i \tag{1}$$

where pi = species proportion, R = total number of species types.

Margalef's Index (MI):

$$MI = \frac{(s-1)}{\ln N} \tag{2}$$

where S = total number of species, N = total number of individuals.

We estimated the biomass of foliage, branches, stems, roots and total tree using an allometric equation relating each biomass component to the diameter at breast height (DBH), respectively. This allometric equation established by Wang (2006) [25].

$$\log_{10}^{B} = a + b(\log_{10}^{DBH}) \tag{3}$$

where B is biomass component, a and b are regression coefficients. The total biomass for trees in each plot was calculated by the biomass of total tree species in the plots.

2.2.2. Soil Sampling and Incubation

The experiment was conducted in May to November 2014 using a modified resin core technique in situ [26–29], similar to the methods of Raison (1987) [30] and Hübner (1991) [31]. At every experimental site, five sampling points were randomly allocated to the replication plots on the first sampling date of 18 May. After the litter and above-ground vegetation was removed, a PVC tube (12 cm long, 5 cm diameter) sharpened in advance was driven 10 cm into the ground to collect the soil core, which was used to determine the initial NO_3^{-}-N, NH_4^{+}-N and mineral N (NH_4^{+}-N and NO_3^{-}-N) concentrations and other soil properties; another identical tube was driven 12 cm into the ground to confine the soil core, with soil structure undamaged, and then the bottom 2 cm of soil was removed, and a resin bag with 10 g anion and cation exchange resin beads (717# and 732# produced by the Huizhi resin Plant of Shanghai) tied into a nylon stocking was placed in the bottom of the PVC tube. The PVC tube containing the soil core and resin was inserted back into its original position and then incubated in situ for one month. At the end of the incubation period, the soil core and resin were removed to determine NO_3^{-}-N, NH_4^{+}-N and mineral N (NH_4^{+}-N and NO_3^{-}-N) concentrations. This was repeated until the experiment ended on 16 November.

2.2.3. Soil Chemical Properties

The collected tubes and soil samples were stored and extracted within 24 h. The resin bags were washed with distilled water and air-dried. The soil samples were homogenized and sieved through a 2-mm screen. For the determination of the soil mineral N (NH_4^{+}-N and NO_3^{-}-N), the soil was shaken with 2 M KCl for 1 h on a 250-rpm shaking table [32]. For the determination of resin NH_4^+-N and NO_3^--N, the resin bags were shaken with 2 M KCl for 12 h on a 250-rpm shaking table and, then the suspension was filtered. The concentrations of NH_4^+-N and NO_3^--N in the extracts were determined by colorimetric methods using a segmented flow injection analyser (Skalar Autoanalyzer SAN^{++}, The Netherlands). The soil microbial biomass C (MBC) and soil microbial biomass N (MBN) were determined by a chloroform fumigation-extraction method [33,34]. Extracts from the fumigated and unfumigated soils (25 g fresh soil) were taken with 50 mL 0.5 M L^{-1} K_2SO_4 for 30 min and filtered. The organic C and N concentrations in the extracts were measured by a dichromate oxidation method and a $K_2S_2O_8$ oxidation method, respectively. MBC and MBN were calculated from the differences in the K_2SO_4-extractable C or N concentration between the fumigated and unfumigated soils divided by the efficiency factors for MBC or MBN ($K_C = 0.38$; $K_N = 0.45$, respectively). K_2SO_4-extractable DOC value was also used as a proxy for the soil available DOC concentration [35]. The soil samples were air-dried and then used for analyses of the soil pH, TN and SOC. The soil pH was determined with a glass electrode (water:soil = 2.5:1). The SOC was analysed using the H_2SO_4-$K_2Cr_2O_7$ oxidation rapid titration method. The TN was measured by the Kjeldahl acid-digestion method. The soil bulk density was measured using the core method. The soil temperature was measured with a thermometer inserted in the soil to a depth of 10 cm.

Nitrogen mineralized was calculated as follows [32]:

$$N_{nit} = [NO_3^- - N]_{sc} + [NO_3^- - N]_{res} - [NO_3^- - N]_{in} \tag{4}$$

$$N_{amm} = [NH_4^+ - N]_{sc} + [NH_4^+ - N]_{res} - [NH_4^+ - N]_{in} \tag{5}$$

$$N_{min} = N_{nit} + N_{amm} \tag{6}$$

where N_{nit}, N_{amm}, and N_{min} are the net nitrification, ammonification and N mineralization, respectively; $[NO_3^- -N]_{sc}$ and $[NO_3^- -N]_{in}$ are the mean concentrations of nitrate N in soil core at the end and beginning of each one month incubation period, respectively; $[NH_4^+ -N]_{sc}$ and $[NH_4^+ -N]_{in}$ are the mean concentrations of ammonium N at the end and beginning of incubation period in soil core, respectively; $[NO_3^- -N]_{res}$ and $[NH_4^+ -N]_{res}$ are the mean concentrations of NH_4^+-N and NO_3^--N in the resin at the end of the incubation period.

2.3. Statistical Analysis

One way analysis of variance (ANOVA) was performed to test the significance of differences in the forest cover characteristics and soil properties, and the least significant difference values (LSD) were calculated at a significance level of $p < 0.05$. The N availability and N mineralization data were tested using the repeated measures analysis of variance (RM-ANOVA). The forest types served as between-subject factors, and months were within-subject factors. The Pearson correlation was used to examine the correlations of mineralization rates with the plant community composition. All statistical analyses were performed with SPSS 14.0 (SPSS Inc., Chicago, IL, USA).

Redundancy analysis (RDA) was used to explore the association of N transformations with environmental properties, using the net organic N mineralization, ammonification and nitrification rates as response variables and the environmental factors as explanatory variables. Before the actual analyses, soil N transformations data were analysed using detrended correspondence analysis (DCA) to determine whether linear or unimodal methods would be appropriate in the analyses [36]. As the eigenvalue was 0.111, redundancy analysis (RDA) was applied. We split the environmental dataset into three groups: climate, plant, and soil variables. In order to avoid high multicollinearity in the following analytical steps we removed within-group correlations of $|r| > 0.7$ (Spearman, $p < 0.05$) by exclusion of variables. By this procedure, selected variables as well as all uncorrelated variables were included in the final matrices. Climate variables include precipitation and temperature; plant variables include biomass of foliage, stems for trees, litter mass for F+H layer, annual litterfall, H' for tree layer and H' for herb layer; soil variables include SOC, DOC, MBC and soil pH. In order to assess the contribution of the three variable groups on N transformation, the variation partitioning method led to the identification of six fractions: independent soil (a), climate (b), plant (c), and joint effects of soil and climate (ab), soil and plant (ac), climate and plant (ac). All multivariate analyses were performed with CANOCO 4.5 [37].

3. Results

3.1. Seasonal Soil Mineral N

NO_3^--N, NH_4^+-N and mineral N contents and dynamics are shown in Figure 2. Generally, the NO_3^--N and mineral N contents in the QJF and QPF plots were higher than those in the LF1 and LF2 plots, and NH_4^+-N contents in the LF2 plots were higher than the QJF plots (Table 2, Figure 2). The mean NO_3^--N contents across the month were 8.87, 7.78, 10.73 and 12.46 mg kg^{-1} in LF1, LF2, QJF and QPF, respectively. The mean NH_4^+-N contents for the LF1, LF2, QJF and QPF were 5.37, 5.73, 4.01 and 5.25 mg kg^{-1}, respectively, and the mean mineral N contents were 14.1, 13.51, 14.74 and 17.71 mg kg^{-1}, respectively. The concentrations of both mineral N and NO_3^--N displayed a distinct seasonal pattern, with generally high concentrations in June and July and low concentrations in August and September (Figure 2). In LF1 and LF2 the maximum soil NH_4^+-N concentration occurred in May, with a second peak value in August. In QJF and QPF, soil NH_4^+-N concentration was higher in May than other months. The ratio of NO_3^--N to NH_4^+-N was approximately 1.79:1 during the growing season, so NO_3^--N was the dominant form of mineral N.

Table 2. Summary of repeated measures ANOVA results on the nitrate N, ammonium N, mineral N, net nitrification rate (N_{nit}), net ammonium rate (N_{amm}) and net mineralization rate (N_{min}).

Factor	NO_3^--N	NH_4^+-N	Mineral N	N_{nit}	N_{amm}	N_{min}
Forest types	54.44 ***	30.31 **	37.54 ***	14.57 **	16.86 **	4.03
Months	94.10 **	263.11 ***	81.71 **	56.26 **	164.12 ***	88.80 ***
Forest types × Months	2.35	15.65 ***	3.54 **	1.02	10.30 ***	1.08

$** p < 0.01, *** p < 0.001.$

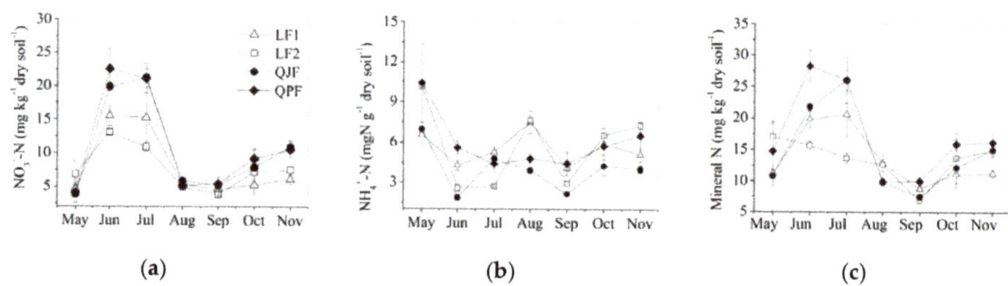

Figure 2. Dynamics of NO_3^--N (a) and NH_4^+-N (b) and inorganic N (c) concentrations in LF1, LF2, QJF and QPF (means ± SE, $n = 3$).

3.2. Net N Mineralization of Organic Matter

The net nitrification rates were significantly different among the months and forest types, but were not different in forest types × months interaction (Table 2). The net nitrification displayed strong seasonal dynamics, with highest rate during the period from June to July and the lowest rate during the period from October to November in LF2 plots (Table 2, Figure 3). In the LF1, QJF and QPF plots, the highest values of net nitrification were observed in the period from May to June. Furthermore, the net N nitrification rates of the QJF and QPF plots were significantly higher than LF2 plots from May to June, and the QPF plots were significantly higher than LF1 from September to October. The mean net N nitrification rates were 8.77, 9.03, 10.86 and 10.81 mg kg^{-1} month^{-1} for LF1, LF2 QJF and QPF, respectively.

The net ammonification rates were significantly different among the months and forest types, and their interactions (Table 2). The net ammonification rate was significantly lower than the rates of nitrification in all of the forest types ($p < 0.01$, Figure 3). In LF1, LF2 and QPF, the net N ammonification rate significantly increased during the periods of May to August and decreased during August to September. In QJF the net N ammonification rate significantly increased during the periods of May to June and decreased during July to September. The mean net N ammonification rates were 2.17, 1.82, 1.13 and 1.44 mg kg^{-1} month^{-1} for LF1, LF2, QJF and QPF, respectively.

The mean net organic N mineralization rates were 10.94, 11.05, 11.25 and 12.25 mg kg^{-1} month^{-1} for LF1, LF2, QJF and QPF, respectively. The net N mineralization displayed strong seasonal dynamics; the temporal variations in the net N mineralization rates were similar to those observed for net nitrification, with the highest rates during the period from June to July and the lowest rates during the period from August to September in all the plots (Figure 3). The net N mineralization rates of the LF1, QJF and QPF plots were significantly higher than LF2 plots from May to June, and the LF2 plots were significantly higher than QJF and QPF from July to August, and the LF2 plots were significantly higher than LF1 from September to October.

The total cumulative nitrification from May to November was significantly higher in QJF and QPF than in LF1 and FL2 ($p < 0.05$, Figure 4a). However, the total cumulative ammonification was lower in QJF and QPF than in LF1 and FL2 ($p < 0.05$, Figure 4b). The total cumulative organic N mineralization was significantly higher in QPF than in LF1 and FL2 ($p < 0.05$, Figure 4c), but QJF was not different with other forest types.

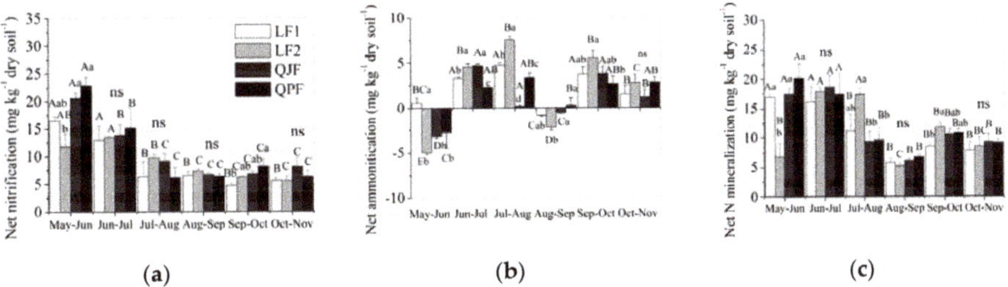

Figure 3. Rates of nitrification (**a**), ammonification (**b**) and net N mineralization (**c**) at one-month intervals (means ± SE, $n = 3$). Different lowercase letters indicate significant differences among forest types; different uppercase letters indicate significant differences among months (adjusted $p < 0.05$).

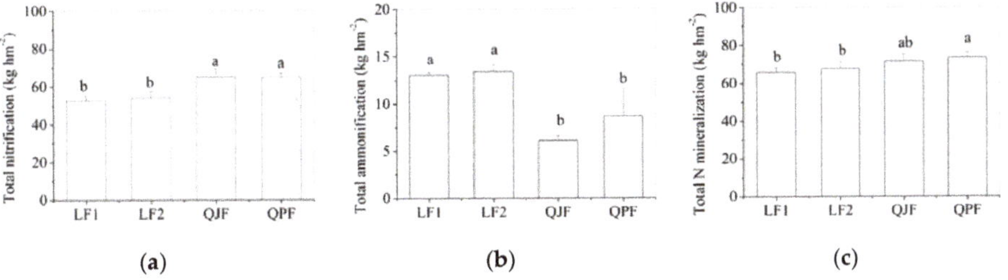

Figure 4. Differences of soil total cumulative nitrification (**a**), ammonification (**b**) and N mineralization (**c**) in various forest types from May to November. Different letters indicate significant differences among forest types (adjusted $p < 0.05$, $n = 3$).

3.3. Plant and Soil Properties

Total biomass estimates of trees in LF2, QJF and QPF were statistically higher than LF1 (Table 3, $p < 0.05$). The biomass recorded in leaves, branches and roots were higher in the QJF than the LF1. The total biomass and below ground biomass for herb layer was not statistically different among forest types (Table 3). Forest floor litter total mass and F+H layer litter mass were highest in the LF2 ($p < 0.05$), and litter mass in L layer was not statistically different among forest types (Table 3). Annual litterfall mass was higher in the QJF and QPF than the LF1 and LF2 (Table 3, $p < 0.05$). The H' and MI were higher in the QJF and QPF than the LF1 and LF2 for tree layer. In herb layer the H' and MI were higher in QPF than other forest types (Table 3, $p < 0.05$). The concentrations of SOC were higher in the QJF than the LF2 (Table 4, $p < 0.05$). The concentrations of TN were higher in the QPF than the LF2 ($p < 0.05$). The MBC concentrations were higher in QPF and QJF than LF1 and LF2 (Table 4, $p < 0.05$). The MBC:MBN ratios were higher in QJF than in LF1 and LF2 (Table 4, $p < 0.05$). The MBN concentrations, C:N ratio, DOC and pH were not statistically different among forest types.

Table 3. Vegetation properties in different forest types (means $\pm$ SE, $n = 3$).

Item	LF1	LF2	QJF	QPF
		Biomass for tree (t/hm^2)		
Foliage	0.71 $\pm$ 0.07c	2.31 $\pm$ 0.43b	3.05 $\pm$ 0.88a	1.96 $\pm$ 0.23b
Branches	4.87 $\pm$ 0.60b	16.28 $\pm$ 3.31ab	34.30 $\pm$ 18.5a	15.10 $\pm$ 2.55ab
Stems	31.20 $\pm$ 1.27c	90.66 $\pm$ 8.67a	83.43 $\pm$ 25.88ab	75.70 $\pm$ 10.52b
Roots	7.29 $\pm$ 0.39c	21.79 $\pm$ 2.69ab	25.20 $\pm$ 7.87a	18.24 $\pm$ 2.32b
Total	44.06 $\pm$ 2.33c	131.03 + 15.04a	145.98 $\pm$ 51.57a	111.01 $\pm$ 15.29b
		Biomass for herb (t/hm^2)		
Above ground	0.93 $\pm$ 0.57c	1.57 $\pm$ 0.21a	1.11 $\pm$ 0.47b	0.96 $\pm$ 0.39c
Below ground	1.81 $\pm$ 0.49a	2.36 $\pm$ 0.59a	2.80 $\pm$ 1.04a	2.27 $\pm$ 0.75a
total	2.75 $\pm$ 1.06a	3.93 $\pm$ 0.66a	3.91 $\pm$ 1.46a	3.23 $\pm$ 0.99a
		Forest floor litter mass (t/hm^2)		
L layer	0.82 $\pm$ 0.16a	0.87 $\pm$ 0.14a	0.77 $\pm$ 0.18a	0.74 $\pm$ 0.23a
F+H layer	3.73 $\pm$ 1.62b	4.46 $\pm$ 0.92a	2.64 $\pm$ 0.50b	2.42 $\pm$ 0.42b
Total	4.53 $\pm$ 1.76b	5.33 $\pm$ 1.05a	3.14 $\pm$ 0.63b	3.61 $\pm$ 0.64b
		Annual litterfall mass (kg/hm^2)		
	30.67 $\pm$ 1.53c	44.64 $\pm$ 9.16b	61.33 $\pm$ 10.21a	62.13 $\pm$ 2.64a
		H' and MI for tree layer		
H'	0.42 $\pm$ 0.07d	0.89 $\pm$ 0.20c	1.31 $\pm$ 0.03b	2.16 $\pm$ 0.25a
MI	0.53 $\pm$ 0.07b	0.63 $\pm$ 0.27b	1.72 $\pm$ 0.24a	1.76 $\pm$ 0.18a
		H' and MI for herb layer		
H'	2.28 $\pm$ 0.13b	2.07 $\pm$ 0.21b	1.99 $\pm$ 0.29b	2.89 $\pm$ 0.54a
MI	4.11 $\pm$ 0.38ab	4.04 $\pm$ 0.96b	3.9 $\pm$ 0.53b	4.93 $\pm$ 0.42a

Different letters indicate significant differences among forest types (adjusted $p < 0.05$).

Table 4. Mean soil properties from May to November in different forest types (means $\pm$ SE, $n = 21$).

Item	LF1	LF2	QJF	QPF
SOC (g/kg dry soil)	60.69 $\pm$ 6.97ab	59.17 $\pm$ 11.94b	66.52 $\pm$ 9.04a	63.22 $\pm$ 5.24ab
TN (g/kg dry soil)	4.99 $\pm$ 0.70ab	4.46 $\pm$ 1.02b	4.90 $\pm$ 1.57ab	5.26 $\pm$ 0.85a
C: N	12.47 $\pm$ 2.91a	10.35 $\pm$ 5.42a	10.40 $\pm$ 5.33a	12.27 $\pm$ 1.94a
MBC (mg/kg dry soil)	403.79 $\pm$ 13.33b	406.81 $\pm$ 27.18b	539.28 $\pm$ 47.51a	543.83 $\pm$ 15.99a
MBN (mg/kg dry soil)	167.66 $\pm$ 38.29a	169.62 $\pm$ 38.80a	176.52 $\pm$ 40.51a	183.30 $\pm$ 30.83a
MBC: MBN	2.54 $\pm$ 0.63b	2.51 $\pm$ 0.52b	3.15 $\pm$ 0.90a	2.97 $\pm$ 0.66ab
DOC (mg/kg dry soil)	129.24 $\pm$ 23.05a	138.07 $\pm$ 50.85a	128.89 $\pm$ 43.84a	135.15 $\pm$ 40.89a
Soil pH	6.39 $\pm$ 0.24a	6.37 $\pm$ 0.24a	6.67 $\pm$ 0.03a	6.41 $\pm$ 0.31a

Different letters indicate significant differences among forest types (adjusted $p < 0.05$).

3.4. Relationship of Soil Organic N Mineralization, Ammonification and Nitrification to Climate, Plant and Soil Variables

3.4.1. Precipitation and Temperature Effects on Net Organic N Mineralization and Nitrification Rates

The net organic N mineralization and nitrification rates in all of the forest types were positively correlated with the precipitation (Figure 5a,c). The net nitrification rates were significantly and positively correlated with precipitation (Figure 5a, $p < 0.01$). In addition, the nitrification rates were positively and significantly correlated with the soil temperature in the LF2 ($p < 0.01$), but not in the LF1, QJF and QPF (Figure 5b). The mineralization rates were significantly and positively correlated with precipitation (Figure 5c, $p < 0.01$). In contrast, the mineralization rates were not significantly correlated with the soil temperature in any forest (Figure 5d).

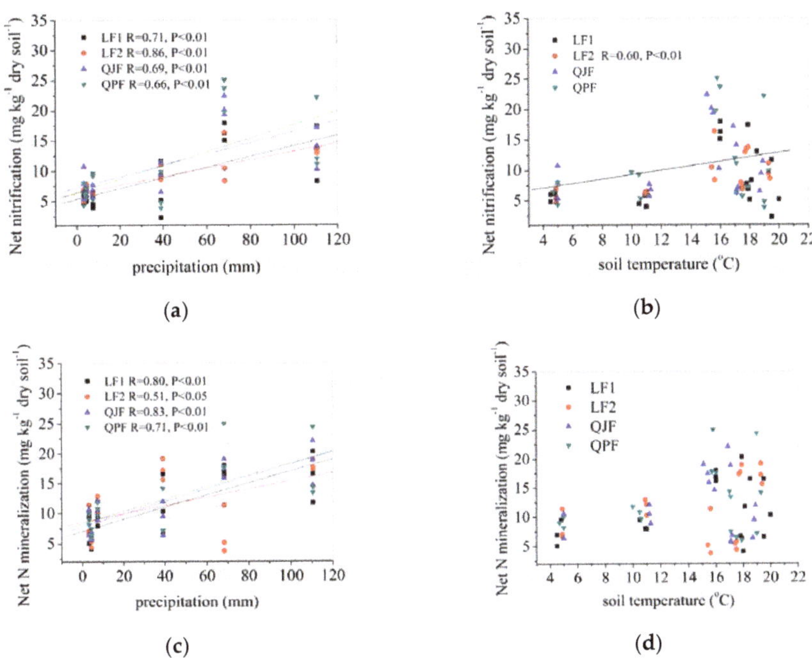

Figure 5. Correlations for nitrification rates to precipitation (**a**) and soil temperature (**b**); correlations for N mineralization rates to precipitation (**c**) and soil temperature (**d**); the precipitation was amount of precipitation per one month and soil temperature was the average for one month.

3.4.2. Correlations between Soil Properties and Organic N Mineralization, Ammonification and Nitrification

The soil net nitrification rate was significantly and positively correlated with the concentrations of MBC ($R = 0.59$, $p < 0.01$), and negatively correlated with soil pH (Figure 6d, $R = -0.71$, $p < 0.01$). There was an increasing trend in soil net nitrification rate with SOC ($R = 0.41$, $p < 0.01$) and DOC ($R = 0.42$, $p < 0.01$) (Figure 6a–d). There was no correlation between the soil net nitrification rate and the TN, MBN and C: N ratio (data not shown). The soil net mineralization rate was significantly and positively correlated with the MBC (Figure 6e, $R = 0.51$, $p < 0.01$). The soil net organic mineralization rate was significantly and negatively correlated with soil pH (Figure 6g, $R = -0.62$, $p < 0.01$). There was no correlation between the soil net N mineralization and the SOC, TN, MBN and C:N ratio (data not shown).

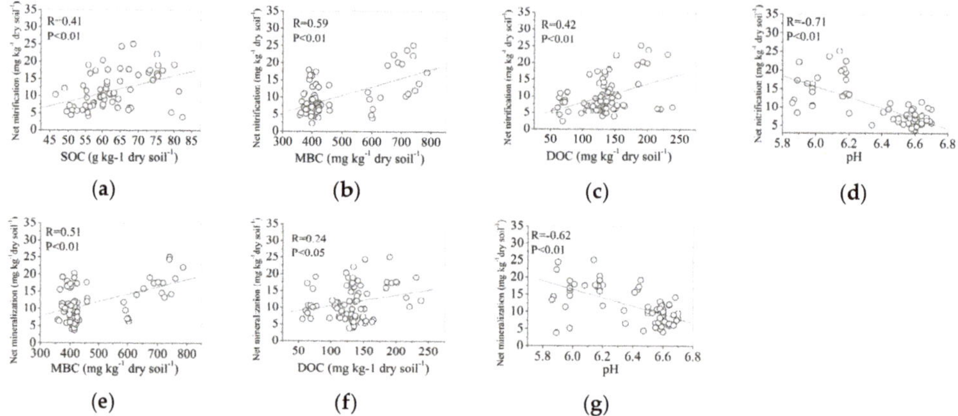

Figure 6. Correlations between net nitrification rates to SOC (**a**), MBC (**b**), DOC (**c**), and pH (**d**), and correlations between net N mineralization rates to MBC (**e**), DOC (**f**), and pH (**g**); the SOC, TN, MBC, DOC and MBC were determined in the beginning of each one-month incubation period, and the data represent the whole experimental period (from May to November) among forest types ($n = 72$).

3.4.3. Relationship between Vegetation Parameters and N Mineralization of Organic Matter

The Pearson correlation was conducted to estimate the relationship between the vegetation parameters and N mineralization of organic matter (Table 5). The results show that soil net nitrification was positively correlated with annual litterfall mass ($p < 0.01$), total foliage biomass ($p < 0.05$), branches biomass ($p < 0.05$), root biomass ($p < 0.01$), H' ($p < 0.01$) and MI ($p < 0.01$) for tree layer. Net ammonification was negatively correlated with annual litterfall mass ($p < 0.01$), total foliage biomass ($p < 0.05$), branches biomass ($p < 0.05$), H' ($p < 0.05$) and MI ($p < 0.01$) for tree layer. Net organic N mineralization was positively correlated with annual litterfall mass ($p < 0.01$), total foliage biomass ($p < 0.05$), H' ($p < 0.01$) and MI ($p < 0.01$) for tree layer and H' for herb layer ($p < 0.05$).

Table 5. Pearson's correlation coefficients (r) between plant variables and soil N mineralization ($n = 12$).

Item	Above Ground	Below Ground	L	F + H	Annual Literfall	Foliage	Branches	Stems	Roots	H'_{tree}	MI_{tree}	H'_{herb}	MI_{herb}
N_{nit}	−0.09	0.53	0.20	−0.32	0.88 **	0.68 *	0.60 *	0.55	0.63 **	0.79 **	0.89 **	0.27	0.42
N_{amm}	0.26	−0.31	0.15	0.51	−0.79 **	−0.61 *	−0.59 *	−0.38	−0.52	−0.65 *	−0.87 *	0.10	−0.17
N_{min}	−0.02	0.57	0.41	−0.14	0.74 **	0.55 *	0.46	0.51	0.53	0.71 **	0.72 **	0.52 *	0.53

$* p < 0.05, ** p < 0.01.$

3.5. Multivariate Analysis (Redundancy Analysis)

To investigate possible relationships between N mineralization of organic matter and environmental variables across forest types and seasons, we performed a Redundancy Analysis (RDA), including organic N mineralization, ammonification, nitrification rates, and the relative climate, plant, and soil variables (Figure 7). A total of 57.6% of variations in seasonal net N mineralization, ammonification, nitrification rates were explained by 13 selected environmental variables. The first two RDA axes explained 49.9% and 7.7% of data variations. Soil nitrification rates were affected by the temperature, precipitation, MBC, H' for tree layer, annual litterfall, total litterfall mass of F + H layer, total foliage and roots biomass for trees, DOC and SOC. Soil N mineralization rates were affected by the temperature, precipitation, MBC and DOC concentration and soil pH. Soil ammonification rates were affected by the total litterfall mass of F + H layer and pH.

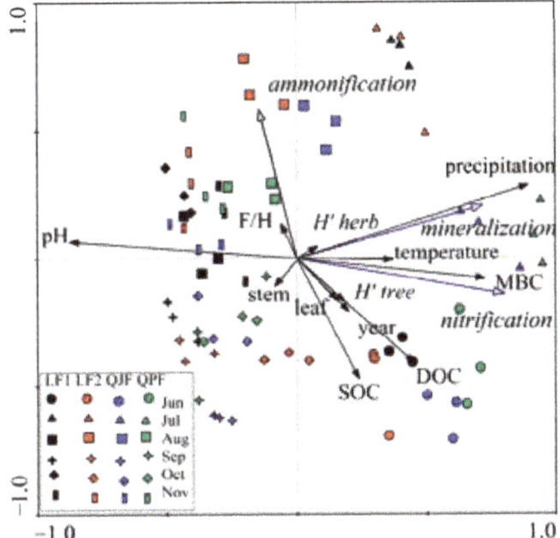

Figure 7. Redundancy analysis using net N transformations data as response variables and climate, plant and soil properties as explanatory variables.

3.6. Variation Partitioning

In order to characterize the relative importance of the broad factors of classification to the soil N mineralization of organic matter in different forest types, variation partitioning was computed independently using climate, soil properties and plant parameters as predictor variables (Figure 8). Results from RDA showed that variables could explain 57.6% of variability in soil N mineralization of organic matter. Both the independent soil (a) and total soil variables (a + ab + ac) accounted for the largest contribution to the variations in soil N mineralization (13.5% and 48.8%), while the independent climatic (b) and total climatic (b + ab + bc) variables were of secondary importance (5.6% and 41.4%). We found 35.8% shared variations of N mineralization explained by soil and/or climatic variables (Figure 7). However, the plant and soil had a negative shared variation of joint fractions (−0.4%).

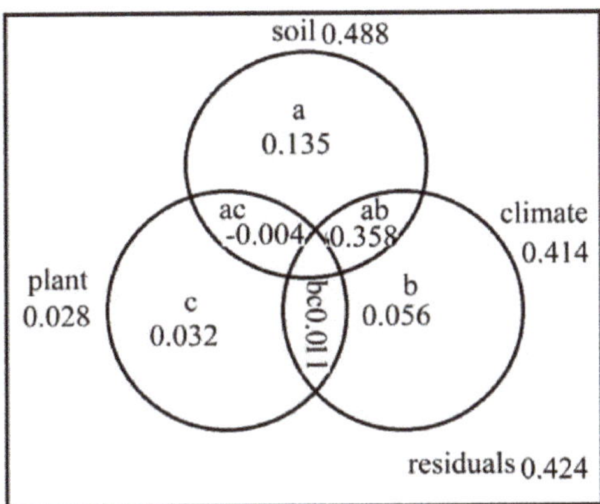

Figure 8. Variation partitioning of N transformations using the matrices of soil variables, plant variables and climate variables. Variables a, b and c denote the independent effects of soil, climate and plant, respectively. Variables ab, ac, bc denote the joint effects of soil and climate, soil and plant, climate and plant, respectively.

4. Discussion

The ranges of net N mineralization of organic matter were different among forest types. Here all of the forest types in this study are considered similar in community structure and historical soil conditions. As such, differences in present soil N mineralization patterns may reflect the impacts of the shifts in plant species composition and plant-induced soil properties on N transformation processes [38,39]. Our results indicate that there were considerable variations among forest types in the plant diversity for tree layer, annual litterfall mass, biomass recorded in leaves, branches, stems and roots, and aboveground biomass for herbs. Surface soil properties were also different among forest types in SOC, TN, MBC and MBC: MBN ratio. We suggest that shifts in plant species composition and resulting plant diversity, forest biomass and soil properties are the most probable explanations for the soil N mineralization patterns of natural secondary forest converted into larch plantations.

4.1. The Effects of Plant on N Mineralization

Our results showed that the two natural secondary forests had higher levels of soil nitrification rates than the two larch plantations, while, net ammonification rates were lower in natural secondary forests than in larch plantations, which could be attributed to the following reasons. Firstly, changes of tree species composition could influence the qualities of litterfall, which may significantly alter the available soil nutrients. Previous work has shown that coniferous tree species typically provide a lower quality of litter material (lower N contents and higher C:N ratio) and a slower litter decomposition, which contribute to a poor soil nutrient level [16,40], and consequently reduced the rate of N cycling [41]. In our study sites, the annual litterfall mass were lower in LF1 and LF2 than QJF and QPF, but the forest floor litter mass was not lower, which indicated the litter decomposition rate was slower in in LF1 and LF2 than QJF and QPF. Secondly, conversion of natural secondary forests to plantations could affect forest biomass productions, which influence the quantities of litterfall. Koutika (2014) found that the higher biomass production may increase soil nitrogen through an enhanced production of litter [42]. In our study, the biomass amounts recorded in leaves, branches, and roots were significantly higher in QJF than LF1 and LF2, thus the annual litterfall mass was higher in QJF than in LF1 and LF2. Our result also indicated soil N mineralization and nitrification rates were positively correlated with the foliage biomass, and the annual litterfall mass. These results suggested that the relatively lower quantities and qualities of litter and slower litter decomposition of larch plantations made N mineralization and nitrification rates decrease after conversion from natural secondary forests. Thirdly, N transformations could also be affected by plant diversity. It has been suggested a more diverse array of plant compositions entering the soil through leaf litterfall could enhance nitrification rates by providing a consistent long-term supply of organic nitrogen as the different qualities of plant material break down at different rates over time. In our study, nitrification rate was positively correlated with plant diversity and richness of tree layer. These results are consistent with previous studies [21,43,44] which also detected a positive plant diversity effect on nitrification.

4.2. The Effects of Soil Properties on N Mineralization

As expected, soil properties were closely correlated with soil organic matter mineralization. Previous studies have shown that soils with high nutrient availabilities could have high N mineralization rates [45,46]. As a biological process, soil N mineralization of organic matter is mainly determined by substrate availabilities and microbial activities [47,48]. We observed the net N mineralization and nitrification rate was positively correlated with MBC, and there was an increasing trend in soil net nitrification rate with SOC and DOC. Soil C and N pools provide available substrates and energy to stimulate microbial activity and which increase the N mineralization rates [39,49]. In our study, MBC concentrations were significantly higher in the QJF and QPF than the LF1 and LF2, the SOC concentrations were significantly higher in the QJF than LF2, TN were significantly higher in the QPF than LF2, consequently, the nitrification was higher in the QJF and QPF than the LF1 and

LF2, the cumulative mineralization was higher in QPF than the LF1 and LF2. Some studies have shown that soil pH is an important factor of soil N mineralization during conversion of broad-leaved forests to coniferous forests. Plantations with coniferous species can produce acidic leaf litters and root exudation of H^+, which in turn lower soil pH and affect micro-fungi activities [50,51]. In our study, soil pH was negatively correlated with net N mineralization and nitrification rate, however, soil pH was not significantly different among forest types, which indicted factors other than pH might have restricted soil N transformation. In this study, we considered the substrate availabilities and microbial activities to be the important factor for N mineralization in organic matter during the forest conversion of natural secondary forests to larch plantations.

4.3. Seasonality Effects on Soil N Transformation Patterns

Seasonal dynamics of temperature and precipitation played an important role in controlling N transformations [11,52]. The present study showed that net nitrification and N mineralization rates were higher from June to July and lower from August to September, whereas the net ammonification rates were higher from July to August and lower from May to June, which can be explained by the seasonal dynamics of temperature and water availability. Soil N mineralization involved biological processes that are moisture and temperature dependent [11,53]. The seasonal changes of temperature and moisture directly control soil microbial activity, which directly influences soil N transformations. In our study, the regression analysis showed that the rates of N mineralization and nitrification had a significantly positive relationship with the precipitation for each forest, while a significant relationship between the net N nitrification and temperature was only found in the LF2. The relationships between the N mineralization rate and precipitation were much stronger than those between the N mineralization rate and temperature, suggesting that precipitation had the major effect on the net N mineralization and nitrification rates in our research region.

4.4. Interaction of Climate, Plant and Soil Variables on N Transformation

In this study, we found a dominant effect of soil properties on N transformations (48.8%). This result is consistent with those of previous studies [10]. Interestingly, the plant variations explained a relatively small fraction in N transformation (2.8%), and the plant and soil had a negative shared variation. This indicated that the interaction effect of plant and soil on N transformation was larger than the summation of independent soil and plant variables. This result can be explained by taking into consideration previous studies wherein plant variables have significant effects on soil properties and which were simultaneously influenced by soil properties, indicating that plant and soil have mutual promoted effects [54]. Climate was also a strong predictor variable of N transformation [8,9]. Many other studies found temperature and precipitation to be the main constraining factors for N transformation [10,11]. In our study, the climate accounted for the secondary contribution to the variations in N mineralization. Overall, these three variations should not be considered mutually exclusive; we may expect each to contribute to the explanation of the potential effects of seasonal dynamics in N mineralization of natural secondary forest converted into larch plantations. In northeast China, plantation/secondary forest landscapes account for the largest proportion of forest areas, and this kind of plantation has been progressively increased throughout the nation. However, we did not study the latitude and longitude, and global climate change effect on N transformation due to limited study sites. Therefore, it is necessary to make a further study on plantation/secondary forest soil N mineralization with a wider study area including latitudinal and longitudinal differences, and varied climatic regions.

5. Conclusions

Our study compared seasonal dynamics of soil N mineralization of organic matter between the natural secondary forests and larch plantations and assessed which could be explained by the plant-soil properties that were associated with forest conversion and seasonal climate conditions.

We demonstrated that plant diversity, litterfall quantity and quality and the soil nutrient availability varied considerably during the forest conversion, which made N mineralization different between the natural forests and plantations. Soil properties were the primary determinant of the N mineralization, and climate conditions also contributed to N mineralization significantly, whereas plant variations were a tertiary contribution to N mineralization. Our results indicated that the larch plantations reduced plant diversity, litter quantity and quality and soil fertility as well as N transformation rate.

Author Contributions: Q.W., F.L., X.R. and Z.F. conceived, designed and installed the experiment; Q.W. and Z.F. were responsible for field work; Q.W. performed laboratory and data analysis under F.L. and X.R. supervision; Q.W. and F.L. wrote the manuscript with contributions from the other authors. All authors approved the final version of the manuscript.

Acknowledgments: This work was supported by the Major Science and Technology Program for Water Pollution Control and Treatment (No. 2012ZX07505-001-01), the National Natural Science Foundation of China (No. 41571464, No. 30972418), and the National Key Technology R&D Program of China (No. 2015BAD07B030102), and the Program of Liaoning Education Department (No. 2017LZD005).

Conflicts of Interest: The authors declare no conflict of interest.

References

1. Pampolino, M.F.; Urushiyama, T.; Hatano, R. Detection of nitrate leaching through bypass flow using pan lysimeter, suction cup, and resin capsule. *Soil Sci.Plant Nutr.* **2000**, *46*, 703–711. [CrossRef]
2. Perakis, S.; Hedin, L. Nitrogen loss from unpolluted south american forests mainly via dissolved organic compounds. *Nature* **2002**, *415*, 416–419. [CrossRef] [PubMed]
3. Pandey, C.B.; Singh, G.B.; Singh, S.K.; Singh, R.K. Soil nitrogen and microbial biomass carbon dynamics in native forests and derived agricultural land uses in a humid tropical climate of india. *Plant Soil* **2010**, *333*, 453–467. [CrossRef]
4. Lupon, A.; Gerber, S.; Sabater, F.; Bernal, S. Climate response of the soil nitrogen cycle in three forest types of a headwater mediterranean catchment: Climate response of soil nitrogen cycle. *J. Geophys. Res. Biogeosci.* **2015**, *120*, 2988–2999. [CrossRef]
5. Kelly, C.N.; Schoenholtz, S.H.; Adams, M.B. Soil properties associated with net nitrification following watershed conversion from appalachian hardwoods to norway spruce. *Plant Soil* **2011**, *344*, 361–376. [CrossRef]
6. Noe, G.B.; Hupp, C.R.; Rybicki, N.B. Hydrogeomorphology influences soil nitrogen and phosphorus mineralization in floodplain wetlands. *Ecosystems* **2013**, *16*, 75–94. [CrossRef]
7. Chen, J.; Xiao, G.; Kuzyakov, Y.; Darrel Jenerette, G.; Ma, Y.; Liu, W.; Wang, Z.; Shen, W. Soil nitrogen transformation responses to seasonal precipitation changes are regulated by changes in functional microbial abundance in a subtropical forest. *Biogeosciences* **2017**, *14*, 2513–2525. [CrossRef]
8. Wang, C.; Wan, S.; Xing, X.; Zhang, L.; Han, X. Temperature and soil moisture interactively affected soil net n mineralization in temperate grassland in northern china. *Soil Biol. Biochem.* **2006**, *38*, 1101–1110. [CrossRef]
9. Wang, L.; Wang, J.; Huang, J. Net nitrogen mineralization and nitrification in three subtropical forests of southwestern china. *Dyn. Soil Dyn. Plant* **2008**, *2*, 33–40.
10. Liu, X.R.; Dong, Y.S.; Ren, J.Q.; Li, S.G. Drivers of soil net nitrogen mineralization in the temperate grasslands in inner mongolia, china. *Nutr. Cycl. Agroecosyst.* **2010**, *87*, 59–69. [CrossRef]
11. Hishi, T.; Urakawa, R.; Tashiro, N.; Maeda, Y.; Shibata, H. Seasonality of factors controlling n mineralization rates among slope positions and aspects in cool-temperate deciduous natural forests and larch plantations. *Biol. Fertil. Soils* **2014**, *50*, 343–356. [CrossRef]
12. Mylliemngap, W.; Nath, D.; Barik, S.K. Changes in vegetation and nitrogen mineralization during recovery of a montane subtropical broadleaved forest in north-eastern india following anthropogenic disturbance. *Ecol. Res.* **2016**, *31*, 21–38. [CrossRef]
13. Chu, H.; Grogan, P. Soil microbial biomass, nutrient availability and nitrogen mineralization potential among vegetation-types in a low arctic tundra landscape. *Plant Soil* **2010**, *329*, 411–420. [CrossRef]
14. Trum, F.; Titeux, H.; Ranger, J.; Delvaux, B. Influence of tree species on carbon and nitrogen transformation patterns in forest floor profiles. *Ann. For. Sci.* **2011**, *68*, 837–847. [CrossRef]

15. Xiong, Y.; Zeng, H.; Xia, H.; Guo, D. Interactions between leaf litter and soil organic matter on carbon and nitrogen mineralization in six forest litter-soil systems. *Plant Soil* **2014**, *379*, 217–229. [CrossRef]

16. Yang, K.; Shi, W.; Zhu, J.J. The impact of secondary forests conversion into larch plantations on soil chemical and microbiological properties. *Plant Soil* **2013**, *368*, 535–546. [CrossRef]

17. Grime, J.P. Benefits of plant diversity to ecosystems: Immediate, filter and founder effects. *J. Ecol.* **1998**, *86*, 902–910. [CrossRef]

18. Khanna, P.K.; Fortmann, H.; Meesenburg, H.; Eichhorn, J.; Meiwes, K.J. Biomass and element content of foliage and aboveground litterfall on the three long-term experimental beech sites: Dynamics and significance. In *Functioning and Management of European Beech Ecosystems*; Brumme, R., Khanna, P.K., Eds.; Springer (Berlin Heidelberg): Berlin, Germany, 2009; pp. 183–205.

19. Denton, C.S.; Bardgett, R.D.; Cook, R.; Hobbs, P.J. Low amounts of root herbivory positively influence the rhizosphere microbial community in a temperate grassland soil. *Soil Biol. Biochem.* **1999**, *31*, 155–165. [CrossRef]

20. Mikola, J.; Barker, G.; Wardle, D. Linking above-ground and below-ground effects in autotrophic microcosms: Effects of shading and defoliation on plant and soil properties. *Oikos* **2000**, *89*, 577–587. [CrossRef]

21. Laughlin, D.C.; Hart, S.C.; Kaye, J.P.; Moore, M.M. Evidence for indirect effects of plant diversity and composition on net nitrification. *Plant Soil* **2010**, *330*, 435–445. [CrossRef]

22. Alarcón-Gutiérrez, E.; Floch, C.; Ziarelli, F.; Albrecht, R.; Le Petit, J.; Augur, C.; Criquet, S. Characterization of a mediterranean litter by ^{13}c cpmas nmr: Relationships between litter depth, enzyme activities and temperature. *Eur. J. Soil Sci.* **2008**, *59*, 486–495. [CrossRef]

23. Shannon, C.E.; Weaver, W. *The Mathematical Theory of Communication*; University of Illinois Press: Urbana, IL, USA, 1949; Volume 27, p. 177.

24. Margalef, R. *Temporal Succession and Spatial Heterogeneity in Phytoplankton*; University of California Press: Berkeley, CA, USA, 1958; pp. 323–349.

25. Wang, C. Biomass allometric equations for 10 co-occurring tree species in chinese temperate forests. *For. Ecol. Manag.* **2006**, *222*, 9–16. [CrossRef]

26. Bhogal, A.; Hatch, D.J.; Shepherd, M.A.; Jarvis, S.C. Comparison of methodologies for field measurement of net nitrogen mineralisation in arable soils. *Plant Soil* **1999**, *207*, 15–28. [CrossRef]

27. Hatch, D.J.; Jarvis, S.C.; Parkinson, R.J.; Lovell, R.D. Combining field incubation with nitrogen-15 labelling to examine nitrogen transformations in low to high intensity grassland management systems. *Biol. Fertil. Soils* **2000**, *30*, 492–499. [CrossRef]

28. Hatch, D.J.; Jarvis, S.C.; Parkinson, R.J. Concurrent measurements of net mineralization, nitrification, denitrification and leaching from field incubated soil cores. *Biol. Fertil. Soils* **1998**, *26*, 323–330. [CrossRef]

29. Carranca, C.; Oliveira, A.; Pampulha, E.; Torres, M.O. Temporal dynamics of soil nitrogen, carbon and microbial activity in conservative and disturbed fields amended with mature white lupine and oat residues. *Geoderma* **2009**, *151*, 50–59. [CrossRef]

30. Raison, R.J.; Connell, M.J.; Khanna, P.K. Methodology for studying fluxes of soil mineral-n in situ. *Soil Biol. Biochem.* **1987**, *19*, 521–530. [CrossRef]

31. Hübner, C.; Redl, G.; Wurst, F. In situ methodology for studying n-mineralization in soils using anion exchange resins. *Soil Biol. Biochem.* **1991**, *23*, 701–702. [CrossRef]

32. Valenzuela-Solano, C.; Crohn, D.M.; Downer, J.A. Nitrogen mineralization from eucalyptus yardwaste mulch applied to young avocado trees. *Biol. Fertil. Soils* **2005**, *41*, 38–45. [CrossRef]

33. Vance, E.D.; Brookes, P.C.; Jenkinson, D.S. An extraction method for measuring soil microbial biomass c. *Soil Biol. Biochem.* **1987**, *19*, 703–707. [CrossRef]

34. Cabrera, M.L. Alkaline persulfate oxidation for determining total nitrogen in microbial biomass extracts. *Soil Sci. Soc. Am. J.* **1993**, *57*, 1007–1012. [CrossRef]

35. Zeglin, L.H.; Stursova, M.; Sinsabaugh, R.L.; Collins, S.L. Microbial responses to nitrogen addition in three contrasting grassland ecosystems. *Oecologia* **2007**, *154*, 349–359. [CrossRef] [PubMed]

36. Lepš, J.; Šmilauer, P. *Multivariate Analysis of Ecological Data Using Canoco*; Cambridge University Press: Cambridge, UK, 2003; p. 269.

37. Borcard, D.; Legendre, P.; Drapeau, P. Partialling out the spatial component of ecological variation. *Ecology* **1992**, *73*, 1045–1055. [CrossRef]

38. Scherer-Lorenzen, M.; Luis Bonilla, J.; Potvin, C. Tree species richness affects litter production and decomposition rates in a tropical biodiversity experiment. *Oikos* **2007**, *116*, 2108–2124. [CrossRef]

39. Lu, S.; Zhang, Y.; Chen, C.; Xu, Z.; Guo, X. Plant–soil interaction affects the mineralization of soil organic carbon: Evidence from 73-year-old plantations with three coniferous tree species in subtropical Australia. *J. Soils Sediments* **2017**, *17*, 985–995. [CrossRef]

40. Fukushima, K.; Tateno, R.; Tokuchi, N. Soil nitrogen dynamics during stand development after clear-cutting of japanese cedar (cryptomeria japonica) plantations. *J. For. Res.* **2011**, *16*, 394. [CrossRef]

41. Burns, D.A.; Murdoch, P.S. Effects of a clearcut on the net rates of nitrification and n mineralization in a northern hardwood forest, catskill mountains, new york, USA. *Biogeochemistry* **2005**, *72*, 123–146. [CrossRef]

42. Koutika, LS.; Epron, D.; Bouillet, J.P.; Mareschal, L. Changes in n and c concentrations, soil acidity and p availability in tropical mixed acacia and eucalypt plantations on a nutrient-poor sandy soil. *Plant Soil* **2014**, *379*, 205–216. [CrossRef]

43. Mulder, C.; Schouten, A.J.; Hund-Rinke, K.; Breure, A.M. The use of nematodes in ecological soil classification and assessment concepts. *Ecotoxicol. Environ. Saf.* **2005**, *62*, 278–289. [CrossRef] [PubMed]

44. Niklaus, P.A.; Kandeler, E.; Leadley, P.W.; Schmid, B.; Tscherko, D.; Körner, C. A link between plant diversity, elevated co2 and soil nitrate. *Oecologia* **2001**, *127*, 540–548. [CrossRef] [PubMed]

45. Buckeridge, K.M.; Zufelt, E.; Chu, H.; Grogan, P. Soil nitrogen cycling rates in low arctic shrub tundra are enhanced by litter feedbacks. *Plant Soil* **2010**, *330*, 407–421. [CrossRef]

46. Zeng, Y.; Xiang, W.; Deng, X.; Fang, X.; Liu, C.; Peng, C. Soil n forms and gross transformation rates in chinese subtropical forests dominated by different tree species. *Plant Soil* **2014**, *384*, 231–242. [CrossRef]

47. Yan, E.R.; Wang, X.H.; Guo, M.; Zhong, Q.; Zhou, W.; Li, Y.F. Temporal patterns of net soil n mineralization and nitrification through secondary succession in the subtropical forests of eastern China. *Plant Soil* **2009**, *320*, 181–194. [CrossRef]

48. Hart, S.C.; Stark, J.M.; Davidson, E.A.; Firestone, M.K. Nitrogen mineralization, immobilization, and nitrification. In *Methods of Soil Analysis, Part 2*; Soil Sci. Society of America, Inc.: Madison, WI, USA, 1994; pp. 985–1019.

49. Taylor, P.G.; Townsend, A. Stoichiometric control of organic carbon-nitrate relationships from soils to the sea. *Nature* **2010**, *464*, 1178–1181. [CrossRef] [PubMed]

50. Jongkind, A.G.; Velthorst, E.; Buurman, P. Soil chemical properties under kauri (agathis australis) in the waitakere ranges, new zealand. *Geoderma* **2007**, *141*, 320–331. [CrossRef]

51. Johnson, D.W.; Miegroet, H.; Lindberg, S.E.; Harrison, R.; Todd, D.E. Nutrient cycling in red spruce forests of the great smoky mountains. *Can. J. For. Res.* **2011**, *21*, 769–787. [CrossRef]

52. Pérez, C.A.; Carmona, M.R.; Aravena, J.C.; Armesto, J.J. Successional changes in soil nitrogen availability, non-symbiotic nitrogen fixation and carbon/nitrogen ratios in southern chilean forest ecosystems. *Oecologia* **2004**, *140*, 617–625. [CrossRef] [PubMed]

53. Yan, E.R.; Wang, X.H.; Huang, J.J.; Li, G.Y.; Zhou, W. Decline of soil nitrogen mineralization and nitrification during forest conversion of evergreen broad-leaved forest to plantations in the subtropical area of eastern china. *Biogeochemistry* **2008**, *89*, 239–251. [CrossRef]

54. Li, Q.; Wang, X.; Jiang, M.; Wu, Y.; Yang, X.; Liao, C.; Liu, F. How environmental and vegetation factors affect spatial patterns of soil carbon and nitrogen in a subtropical mixed forest in central china. *J. Soils Sediments* **2017**, *17*, 2296–2304. [CrossRef]

 *forests*

Article

Thinning Treatments Reduce Deep Soil Carbon and Nitrogen Stocks in a Coastal Pacific Northwest Forest [†]

Cole D. Gross [1],*, Jason N. James [2], Eric C. Turnblom [2] and Robert B. Harrison [2]

[1]   Department of Renewable Resources, University of Alberta, 442 Earth Sciences Building, Edmonton, AB T6G 2E3, Canada

[2]   School of Environmental and Forest Sciences, University of Washington, Box 352100, Seattle, WA 98195-2100, USA; jajames@uw.edu (J.N.J.); ect@uw.edu (E.C.T.); robh@uw.edu (R.B.H.)

*   Correspondence: cgross@ualberta.ca

†   This paper is included as a chapter of the Master Degree Thesis of the first author entitled Soil Carbon and Nitrogen Stocks: Underestimation with Common Sampling Methods, and Effects of Thinning and Fertilization Treatments in a Coastal Pacific Northwest Forest published by the University of Washington in 2017.

Received: 4 April 2018; Accepted: 28 April 2018; Published: 1 May 2018

Abstract: Forests provide valuable ecosystem and societal services, including the sequestration of carbon (C) from the atmosphere. Management practices can impact both soil C and nitrogen (N) cycling. This study examines soil organic C (SOC) and N responses to thinning and fertilization treatments. Soil was sampled at an intensively managed Douglas-fir (*Pseudotsuga menziesii* (Mirb.) Franco) plantation in north-western Oregon, USA. Management regimes—thinning, fertilization plus thinning, and no (control) treatment—were randomly assigned to nine 0.2-ha plots established in 1989 in a juvenile stand. Prior to harvest, forest floor and soil bulk density and chemical analysis samples were collected by depth to 150 cm. During a single rotation of ~40 years, thinning treatments significantly reduced SOC and N stocks by 25% and 27%, respectively, compared to no treatment. Most of this loss occurred in deeper soil layers (below ~20 cm). Fertilization plus thinning treatments also reduced SOC and N stocks, but not significantly. Across all management regimes, deeper soil layers comprised the majority of SOC and N stocks. This study shows that: (1) accurately quantifying and comparing SOC and N stocks requires sampling deep soil; and (2) forest management can substantially impact both surface and deep SOC and N stocks on decadal timescales.

Keywords: soil organic carbon; carbon sequestration; nitrogen; deep soil; forest floor; forest management; fertilization; thinning; fixed depth; equivalent soil mass

1. Introduction

The world's forests are an important terrestrial carbon (C) sink, sequestering as much as 30% (~2 Pg C y^{-1}) of annual global anthropogenic CO_2 emissions between 1990 and 2007 [1,2]. In addition to their importance in the global C cycle, forests provide many other valuable ecosystem and societal services. Forest-management practices can enhance or reduce the ability of a given forest stand to act as a C sink and provide these services [3,4]. Since approximately two-thirds of forests are managed [1], understanding how forest-management practices affect the global C cycle and the capacity to sustainably produce natural resources is a high priority.

While much research has been conducted regarding the aboveground effects of forest management, comparably little is known about the effects belowground [5]. Soils comprise the majority of the terrestrial C stock [6] and account for ~85%, 60%, and 50% of the total C stock in boreal forest, temperate forest, and tropical rainforest ecosystems, respectively [3,7]. Therefore, determining the fate of forest soil organic C (SOC) in response to management is an essential part of understanding climate-carbon feedbacks and changes in forest ecosystem C budgets. Gains or losses in SOC affect numerous soil properties essential to maintaining beneficial ecosystem services and productive forest stands, including the water- and nutrient-holding capacity of the soil [8].

In Pacific Northwest forest ecosystems, nitrogen (N) is often the primary limiting nutrient [9–11]. The fate of soil N in response to forest management is thus a key concern in this region. In general, N fertilization stimulates biomass production, but the effects on SOC and other soil-nutrient dynamics are variable and highly site dependent due to complex interactions between soil properties, microorganisms and vegetation [12]. Increases in soil N have the potential to enhance aboveground biomass growth and forest productivity, as well as to increase SOC and the retention of other nutrients in the soil through a combination of increased belowground biomass and delayed root decomposition [13].

In a recent meta-analysis, James and Harrison [14] found that harvesting reduced soil C by an average of ~11% globally. Significant losses in both the litter layer (O horizon) and the mineral soil were observed, with particularly large losses in very deep soil (60–100+ cm in depth) compared to more superficial soil. Interestingly, thinning treatments resulted in greater losses of mineral soil C than clear-cut harvesting by a difference of ~9% [14]. Although this may seem counterintuitive, thinning treatments can lead to less SOC accumulation over time due to reduced root C inputs (from reduced root biomass) [15], and they can also affect numerous soil properties that enhance microbial metabolic activity or encourage increased leaching and export of SOC and other soil nutrients. Mechanisms that can lead to SOC and N losses due to thinning include: (1) increased soil temperature; (2) microbial stimulation or priming; (3) nitrate leaching; and (4) groundwater-level rise.

1.1. Increased Soil Temperature

Decreased radiation interception by trees due to thinning treatments can result in soil temperature increases [16]. Several studies examining the upper 10 cm of soil have shown post-thinning increases in soil temperature ranging from 1–6 °C when compared to reference stands [16–18]. Soil temperature also tends to increase with thinning intensity [16,17]. Along with increases in soil temperature, Cheng et al. [17] measured an increase in soil respiration ranging from ~8% to 20% and increasing with thinning intensity. Hicks Pries et al. [19] found that mineral soil respiration in a temperate forest ecosystem increased by 34–37% to a depth of 100 cm when subjected to 4 °C warming, with soil below 15 cm contributing to ~50% of the total respiration. Forest harvest has been observed to increase mean soil temperature and mean daily soil temperature flux by ~3 °C and 5 °C, respectively, at 10 cm in depth and by ~2 °C and 3 °C, respectively, at 100 cm in depth [20]. Any increase in SOC decomposition rates also increases microbial demand for N, as microorganisms require about a 24:1 C:N ratio during organic matter decomposition [21]. Competition for available N in the soil environment is increased in particular during the decomposition of organic matter with a higher C:N ratio, such as coarse woody debris left as slash on the forest floor post-thinning, and can lead to N scavenging as plants and microorganisms compete for this essential nutrient [21,22].

1.2. Microbial Priming

Microbial priming is a mechanism by which potential energetic barriers to SOC decomposition are alleviated by the introduction of fresh C compounds [23]. Similarly, higher rates of nitrification and additional nitrate can result in increased SOC decomposition and dissolved organic carbon (DOC) production by alleviating microbial nutrient limitations [24,25]. The priming phenomenon is particularly relevant in deeper soil layers where SOC is often thousands to tens of thousands of years old [23,26]. When the environmental conditions under which deep SOC accumulated change, such as through the addition of fresh C compounds, this SOC is vulnerable to decomposition [19,23,26–28]. Removing trees results in decreased transpiration and rain interception, and in turn can lead to increased DOC flux and transport to deeper soil layers, which is driven largely by rain events and new inputs of organic matter [29,30]. Organic matter left on-site post-thinning (e.g., roots and slash) substantially increases soil C inputs and the potential for priming effects. While forest floor fresh C inputs are often mineralized in the litter layer [26,30], C inputs due to the mineralization of root biomass can persist for many years following the harvest of trees [31]. Post-thinning increases in channels of decaying roots, particularly coarse roots, could also create preferential flow paths for DOC transport. Along these pathways, DOC has fewer opportunities for abiotic and biotic interactions, potentially introducing large amounts of fresh DOC to deeper soil layers [32]. As preferential pathways have been shown to have greater SOC concentrations and microbial biomass than the surrounding bulk soil [33], mineralization rates in these pathways are likely enhanced [34], which may lead to increased SOC and N losses from the adjacent bulk soil.

1.3. Nitrate Leaching

Several studies have observed increased leaching or export of nitrate (and other forms of N) following forest harvest under logging residues [35–39]. Rosen and Lundmark-Thelin [39] attributed this phenomenon to a combination of reduced N uptake by roots and increased mineralization of the litter layer. Fertilized forest stands may be particularly vulnerable to nitrate leaching under slash left on the forest floor post-thinning due to increased N availability. High rates of nitrification, nitrate leaching, and increased soil acidity due to fertilization can negatively affect soil quality through the priming of SOC and through the co-leaching of other nutrients such as calcium and magnesium [40].

1.4. Groundwater-Level Rise

The removal of trees via thinning treatments or harvest reduces transpiration, which can result in groundwater-level rise and potential increases in the export of DOC and various forms of N [37,41]. Laudon et al. [41] observed a >70% increase in DOC export from harvested compared to unharvested sites one year post-harvest. They attributed this increase in DOC export primarily to a raised groundwater level contacting more superficial soil layers that have higher DOC (and N) concentrations. Depending on the original height of the water table prior to the removal of trees, substantial portions of deeper SOC and N could be especially vulnerable to export and loss via this mechanism.

1.5. Summary and Objectives

Jandl et al. [12] concluded that, in general, forest thinning increases the stability of a stand at the expense of SOC stocks. However, other studies have found no difference in SOC stocks post-thinning [5]. In a long-term reforestation study in the subtropical south-eastern USA. Mobley et al. [42] found that thinning treatments reduced SOC and N stocks in particular in deeper soil layers where losses exceeded new inputs. Although deeper soil is often ignored in short- and even long-term studies, the observation of deep SOC losses on decadal timescales due to management or land-use change is not uncommon [42–44]. Considering that most SOC is contained in deeper soil layers (below ~20 cm) [6,45,46], and most trees root deeper than 100 cm [47], studying deep-soil nutrient dynamics is essential to understanding forest-management effects on SOC stocks. Unfortunately, soils are often sampled to 20 cm or less and are rarely sampled below 100 cm [4,6,14,27,42,45,48].

The objective of this study was to determine the responses of SOC and N to thinning and fertilization plus thinning treatments of varying intensity, and to understand how any observed response differed vertically in the soil profile to a depth of more than 100 cm. Both the fixed-depth and mass-based approaches were used to quantify and compare SOC and N stocks. We found that thinning treatments substantially reduced both SOC and N stocks, particularly in deeper soil layers, highlighting that forest-management practices can affect both surface and deep SOC and N stocks on decadal timescales, and that accurately quantifying and comparing SOC and N stocks requires sampling deep soil.

2. Materials and Methods

Soil was sampled at an intensively managed Douglas-fir (*Pseudotsuga menziesii* (Mirb.) Franco) plantation in north-western Oregon, USA (Figure 1, inset). The plantation was planted on commercial forest land in 1977 with 2-year-old Douglas-fir seedlings and was ≥90% Douglas-fir throughout the ~40-year rotation. The previous plantation, also Douglas-fir, underwent uniform treatment and harvest. Square plots were established in 1989, the boundaries of which were designed to maximize uniformity within and between plots. Management regimes—thinning (T_{trt}), fertilization plus thinning (FT_{trt}), and no (control) treatment (C_{trt})—were randomly assigned to nine 0.2-ha plots spanning a total area of ~5 ha (Figure 1). Three plots received thinning treatments; three plots received fertilization plus thinning treatments of varying intensity; and three plots received no (control) treatment. Additionally, the initial trees per hectare were systematically reduced to one-half or one-fourth on randomly assigned plots (Table 1). The removed trees were left on the forest floor. Fertilized plots received 224 kg N ha^{-1} as urea every 4 years starting in 1989 for a total of 1120 kg N ha^{-1} over 16 years. Thinning treatments were based on Curtis' [49] relative stand density. The stems of thinned trees were either removed or left on the forest floor, depending on the size of the trees at the time of treatment. Slash from trees was consistently left on the forest floor, even when stems were removed. Landform across all plots was nearly level to gently sloping, with an average slope of 10% and a maximum slope of <30%. The climate is characterized by cool, wet winters and warm, dry summers. Mean annual precipitation from 1981 to 2010 was ~220 cm, and mean annual temperature was 9 °C, with an annual maximum and minimum temperature of 14 °C and 4 °C, respectively [50]. Elevation ranged from 620 m to 660 m above sea level. Soil was moderately well drained with a low mean rock fragment content by sample weight (<2% fine to medium gravel). The soil sampled was an older, clayey soil (a Palehumult, closely resembling the Cumley series), making it reasonably uniform and an ideal soil for sampling to compare forest-management treatments.

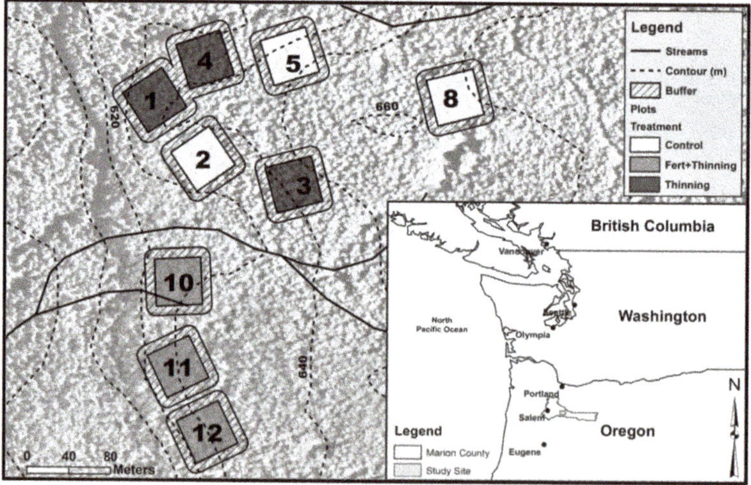

Figure 1. Location of study site (inset) and layout of treatment plots. The site coordinates are 44.87417, −122.566 (Latitude/Longitude decimal degrees). Fert, fertilized.

Table 1. Summary of plot treatments. The plantation was planted in 1977 with 2-year-old Douglas-fir seedlings. Plots are each 0.2-ha and span a total area of ~5 ha. Fert, fertilized.

Plot	Initial Trees (ha^{-1})	Treatment Group	Fertilized [§]	Thinning Treatment	Thinning Year	Stems Left on Forest Floor [#]
1	1117	Thinning	No	RD55→RD35	2001	Yes
2	297 [†]	Control	No	None	-	-
3	558 [‡]	Thinning	No	RD55→RD35	2011	No
4	1181	Thinning	No	RD55→RD35	1999	Yes
5	1196	Control	No	None	-	-
8	554 [‡]	Control	No	None	-	-
10	1240	Fert + Thinning	Yes	RD55→RD35	1997	Yes
				RD55→RD40	2005	No
11	558 [‡]	Fert + Thinning	Yes	RD55→RD35	2009	No
12	311 [†]	Fert + Thinning	Yes	None	-	-

[†] initial trees per hectare were reduced to one-fourth in 1989; stems and slash left on forest floor. [‡] initial trees per hectare were reduced to one-half in 1989; stems and slash left on forest floor. [§] fertilized plots received 224 kg N ha^{-1} as urea every four years starting in 1989 for a total of 1120 kg N ha^{-1} over 16 years. RD = Curtis's [49] relative density. [#] slash consistently left on forest floor, even when tree stems were removed.

Three soil pits per plot were excavated with a shovel to 100 cm or 150 cm. Major genetic horizons, soil colors, textures and structures were identified, and roots and stone content were recorded (Table 2). Soil bulk density and chemical analysis samples were collected in the late summer and early fall of 2015, immediately preceding harvest. Soil was collected from all nine plots over (three) two-week time periods to minimize any differences in weather conditions between plots during the sampling timeframe. Samples were collected randomly from within the middle of soil depth layers 0–10, 10–20, 20–50, 50–100, and 100–150 cm. One forest floor sample was collected from a randomly placed 20 × 30 cm quadrat nearby each soil pit. All soil samples were analyzed separately, and repeated measurements within a plot and soil depth layer were subsequently averaged to account for within-plot variation. Soil bulk density and chemical analysis samples were collected using a 5.4-cm diameter hammer-core, as well as using clod and excavation (irregular hole, water replacement) methods. Repeated measurements within the same plot and soil depth layer across methods were analyzed separately to adjust for differences between the methods using regression (wherein the excavation method was used as the standard). A detailed description of soil-sampling techniques, laboratory methods, and regression equations is found in Gross and Harrison [51]. All soil samples collected in the field were sealed in plastic bags, returned to the laboratory within 48 h, and stored at 3 °C until analysis.

Soil subsamples used in elemental analysis were taken from bulk density samples to avoid potential biases, as SOC concentrations and bulk density are not independent variables [52]. Air-dried samples were sieved to <4.75 mm (rather than to <2 mm) to avoid discarding a meaningful portion of SOC [53,54]. The >4.75 mm fraction was weighed and the volume determined by displacement of water in a graduated cylinder. Litter layer samples were weighed, air-dried to a constant weight, reweighed, and ground to less than ~0.5 mm. Representative subsamples of litter layer and <4.75 mm mineral soil fractions were ground with a mortar and pestle and analyzed for total C and N concentrations (g kg^{-1}) using an automated elemental analyzer (Perkin-Elmer 2400, PerkinElmer, Waltham, MA, USA). Approximately 20% of the samples were run twice to verify the precision of the analysis, and quality control samples of known C concentration were run every 10 samples. The average of samples run twice was used for analysis. Due to a lack of carbonates measured in the region [55] and strongly acid soils (pH < 6), total C concentrations are equated to organic C [56].

Table 2. Typical soil profile description. Resembles Cumley series (a Palehumult). Average slope of 10%. Mean elevation ~640 m above sea level. Gr, gravel (0.2–7.5 cm); Cb, cobbles (7.5–25 cm); St, stones (>25 cm); YR, yellow-red.

Horizons	Depth (cm)	Color (moist)	Texture	Structure	Roots	Rock (%)
Oi	3.5–0					Gr < 2% Cb < 5% St < 5%
A	0–15	10 YR 3/3 dark brown	Sandy clay loam	Medium to coarse granular, weak to moderate	Many fine, medium, and coarse	Gr < 2% Cb < 5% St < 5%
BA	15–30	10YR 3/4 dark yellowish brown	Sandy clay	Medium to coarse subangular blocky, moderate	Many fine and medium; few coarse	Gr < 2% Cb < 5% St < 5%
Bt1	30–80	5YR 4/4 reddish brown	Sandy clay to clay	Coarse to very coarse subangular blocky, moderate to strong	Common fine, few medium; very few coarse	Gr < 2% Cb < 5% St < 5%
Bt2	80–150+	10YR 4/4 dark yellowish brown	Sandy clay	Coarse subangular blocky, moderate	Few fine and medium; very few coarse	Gr < 2% Cb < 5% St < 5%

Soil pH was measured in a 1:1 (deionized H_2O, mL; soil, g) mix for mineral soil and a 4:1 mix for the litter layer with a digital pH meter (Model PC-700, Oakton Inst., Vernon Hills, IL, USA). Soil mixtures were stirred and left to stand undisturbed for at least 30 min to allow homogenization before pH was measured. The volume of field-moist clods [57] was determined by the paraffin wax method and displacement of water. Oven-dry weights for the clod method were determined by drying the clods in the oven at 105 °C for at least 48 h. Subsamples of each core and excavation sample were dried for at least 48 h at 105 °C, and oven-dry weights were determined by applying conversions to the air-dried weights. Bulk density was calculated according to:

$$\text{Bulk density} = \left(\frac{\text{oven dry sample weight, g} - \text{rock fragment weight, g } (> 4.75 \text{ mm})}{\text{soil volume, cm}^3 \text{ (solids} + \text{pores)} - \text{rock fragment volume, cm}^3 (> 4.75 \text{ mm})} \right), \tag{1}$$

Soil organic C and N stocks were determined using the fixed-depth equations:

$$\text{Mg SOC ha}^{-1} = \left(\frac{\text{mg SOC}}{\text{g soil}} \right) \left(\frac{\text{g soil}}{\text{cm}^3 \text{ soil}} \right) \left(\frac{\text{cm}}{1} \right) \left(\frac{\text{Mg}}{10^9 \text{ mg}} \right) \left(\frac{10^8 \text{ cm}^2}{\text{ha}} \right), \tag{2}$$

and,

$$\text{kg N ha}^{-1} = \left(\frac{\text{mg N}}{\text{g soil}} \right) \left(\frac{\text{g soil}}{\text{cm}^3 \text{ soil}} \right) \left(\frac{\text{cm}}{1} \right) \left(\frac{\text{kg}}{10^6 \text{ mg}} \right) \left(\frac{10^8 \text{ cm}^2}{\text{ha}} \right), \tag{3}$$

where mg SOC g soil^{-1} is SOC concentration, mg N g soil^{-1} is N concentration, g soil cm soil^{-3} is bulk density, and cm is soil layer thickness. The mass-based approach according to the procedure of Wendt and Hauser [58] was also used to estimate SOC and N stocks. For this approach, soil-sample mass for each depth layer ($M_{SAMPLE(DL)}$, g) and method was calculated according to:

$$M_{SAMPLE(DL)} = \pi \left(\frac{5.4 \text{ cm}}{2} \right)^2 \times \left(\frac{\text{soil layer thickness, cm}}{1} \right) \times \text{bulk density}, \tag{4}$$

where 5.4 cm is the inside diameter of the core. Soil-sample masses, SOC and N concentrations, the inside diameter of the core, and the number of cores per sample were subsequently input into the web-accessible spreadsheet [59] created by Wendt and Hauser [58], which fits a cubic spline function to model the relationship between cumulative areal soil mass and cumulative SOC mass. Reference mass layers were set using the lowest mean soil mass across treatments for each soil depth layer as recommended by Lee et al. [60] for systems in which the initial conditions (e.g., SOC or bulk density) are not available.

In this study, both the fixed-depth and mass-based approaches used to calculate SOC and N stocks replace the volume of the >4.75 mm fraction with fine soil (<4.75 mm). Multiple studies have found that this method has the potential to overestimate [61] or underestimate [27,54,62–64] SOC and N stocks. However, these errors appear to be limited to rocky soils [54,61,62,64]. Because the soil sampled in the current study was generally non-rocky (<2% fine to medium gravel content by weight), this method is unlikely to cause substantial biases in SOC and N stock estimates.

Sampling in the middle of a soil depth layer likely underestimates SOC and N concentrations and overestimates the bulk density of soil above the sample, while resulting in errors in the opposite direction concerning soil below the sample (i.e., overestimating SOC and N concentrations and underestimating bulk density). An assumption of this sampling method is that the errors tend to have a canceling effect, giving an accurate estimate of SOC and N stocks in the given layer. This assumption applies to all calculations of mineral SOC and N stocks in this study. As the entire depth of the litter layer was sampled, this assumption does not apply to the litter layer. Additionally, litter layer SOC and N stocks were only determined using Equations (2) and (3).

Total aboveground and root C sequestration were determined using the biomass ratio equation and parameters provided by Table 6 in Jenkins et al. [65]. For a given plot, the quadratic mean diameter (QMD) was substituted for diameter at breast height (DBH). Total aboveground biomass (TAB) for each plot was calculated according to:

$$\text{TAB Mg ha}^{-1} = \frac{\left(\frac{450 \text{ kg}}{\text{m}^3}\right)\left(\frac{\text{stand volume, m}^3}{\text{ha}}\right)\left(\frac{\text{Mg}}{1000 \text{ kg}}\right)}{e^{\left(-0.3737 - \left(\frac{1.8055}{\text{QMD,cm}}\right)\right)}},$$

(5)

where 450 kg m^{-3} is the density of Douglas-fir [66], and the denominator is the biomass ratio equation for softwood-stem wood. The sum of the removed and final aboveground biomass was used for TAB for thinned plots to represent total aboveground and root C sequestration. A ratio of 0.512, which is the average C concentration (g C g^{-1}) for Douglas-fir trees in the Pacific Coast and Rocky Mountain regions [67], was used to convert total mass to C mass. Belowground C stocks were determined by summing the SOC stock (calculated using the mass-based approach) for a given plot with total root C for the same plot. Total C stocks were determined by summing the above- and belowground C stocks for a given plot. Merchantable volume for thinned plots was calculated by including removed volume only if the given volume was commercially thinned.

Analysis of variance (ANOVA) was used to determine whether treatment affected various soil properties and SOC and N stocks. When significant differences were detected at $p < 0.1$, Tukey's honest significant difference (HSD) post-hoc tests were conducted to compare the means. A significance level of 0.1 was chosen (rather than 0.05, for example) to reduce the chance of false negatives [68], as forest-management effects on soil properties or nutrient stocks can have lasting impacts on soil health and future productivity. Relationships between numerical variables such as SOC and N concentrations were determined using linear regression. Data met the assumptions of the analyses performed and no data transformations were necessary. Data were analyzed using R studio [69].

3. Results

3.1. Soil Bulk Density and Organic Carbon Concentrations

In the mineral soil layers, bulk density was consistently highest for T_{trt} at all depths and lowest for C_{trt} to a depth of 100 cm (Figure 2). In the upper 50 cm, this pattern was inversely related to SOC concentrations, with C_{trt} having the highest SOC concentrations and T_{trt} having the lowest (Figure 2). Soil organic C concentration was a significant ($p < 0.1$) predictor of bulk density in the mineral soil, the two variables following a quadratic relationship (Figure 3). However, a large amount of variation occurred where SOC concentrations were below ~15 g SOC/kg soil, which corresponded to depths below 50 cm. When the relationship between SOC concentration and bulk density was analyzed by depths 0–50 cm and 50–150 cm, the former explained slightly more of the variation in bulk density values than the whole mineral soil (i.e., 0–150 cm in depth), while the latter was not significant (Figure 3). Soil organic C concentration and bulk density were not significantly related in the litter layer (Figure 3). Bulk density for T_{trt} was significantly (Tukey's HSD, $\alpha = 0.1$) higher than C_{trt} in the upper 20 cm of mineral soil and significantly higher than FT_{trt} in the 10–20 cm depth layer. C_{trt} had a significantly higher SOC concentration compared to both T_{trt} and FT_{trt} in the upper 20 cm of mineral soil, as well as in the 100–150 cm depth layer compared to FT_{trt}. No significant differences between treatment means were observed for bulk density or SOC concentration in the litter layer.

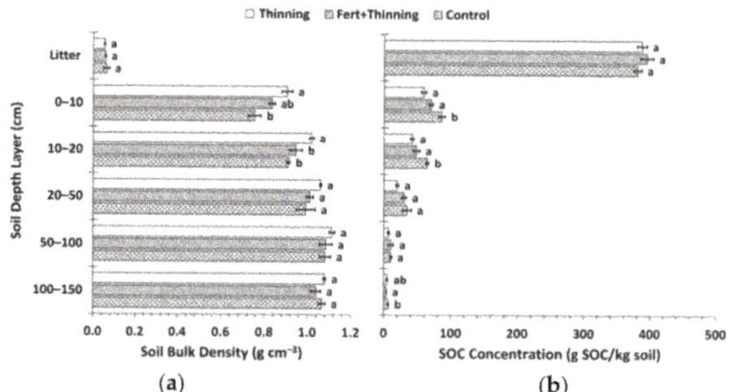

Figure 2. (**a**) Mean soil bulk density by treatment; (**b**) mean soil organic carbon (SOC) concentration by treatment. Sample size for all treatments and soil depth layers is three. Error bars represent ± one standard error. Means within each soil depth layer accompanied by the same letter (i.e., "a" or "b") are not significantly different (Tukey's honest significant difference (HSD), $\alpha = 0.1$). Fert, fertilized.

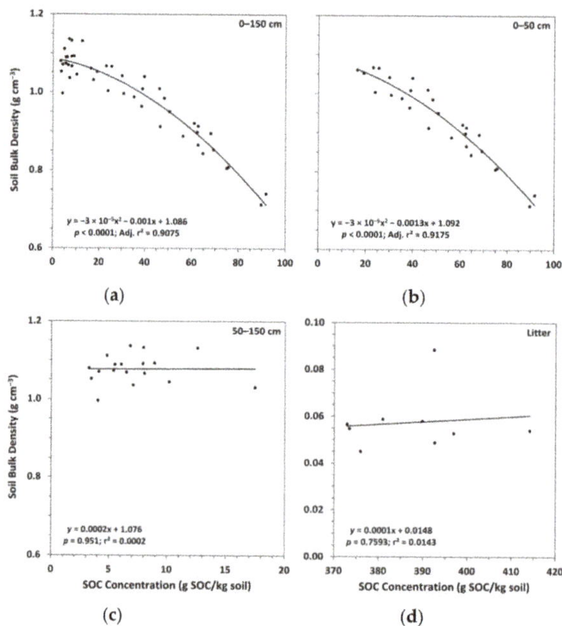

Figure 3. Soil bulk density versus soil organic carbon (SOC) concentration for: (**a**) the whole mineral soil ($n = 45$); (**b**) the 0–50 cm soil depth layers ($n = 27$); (**c**) the 50–150 cm soil depth layers ($n = 18$); (**d**) the litter layer ($n = 9$).

3.2. Soil Nitrogen Concentrations, Carbon to Nitrogen Ratios, and pH

Trends in soil N concentrations between treatments in the mineral soil followed a similar pattern as for SOC concentrations, although a few additional means were found to be significantly different (Figure 4). C_{trt} had a significantly higher soil N concentration compared to T_{trt} at all depths except the 50–100 cm depth layer. FT_{trt} also had a significantly higher soil N concentration compared to T_{trt} in the 0–10 cm depth layer. Compared to FT_{trt}, C_{trt} had a significantly higher soil N concentration in the upper 20 cm of mineral soil and in the 100–150 cm depth layer. Notably, in the litter layer, T_{trt} had a significantly lower soil N concentration compared to the other two treatments. No significant differences in the C:N ratio (SOC concentration/N concentration) in the mineral soil were observed (Figure 4). In the litter layer, T_{trt} had a significantly higher C:N ratio compared to C_{trt} and FT_{trt} (39, 32, and 33, respectively). Surface soil layers (0–20 cm in depth) had a mean C:N ratio of ~24 across all treatments, while deeper soil layers had normalized mean C:N ratios of 18, 19, and 17 for T_{trt}, FT_{trt}, and C_{trt}, respectively. Soil organic C concentration was a significant predictor of N concentration in the mineral soil, the two variables being positively related, and explained much of the variation in N concentration ($r^2 = 0.9886$) (Figure 5). These two variables were not significantly related in the litter layer (Figure 5).

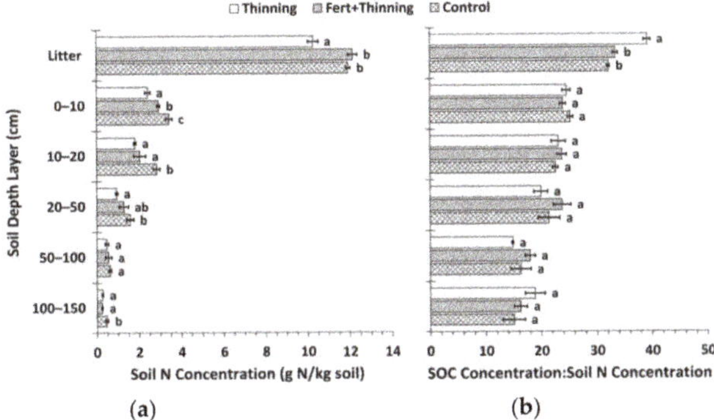

Figure 4. (a) Mean soil nitrogen (N) concentration by treatment; (b) mean soil organic carbon (SOC) concentration to N concentration by treatment. Sample size for all treatments and soil depth layers is three. Error bars represent ± one standard error. Means within each soil depth layer accompanied by the same letter (i.e., "a" or "b") are not significantly different (Tukey's HSD, $\alpha = 0.1$). Fert, fertilized.

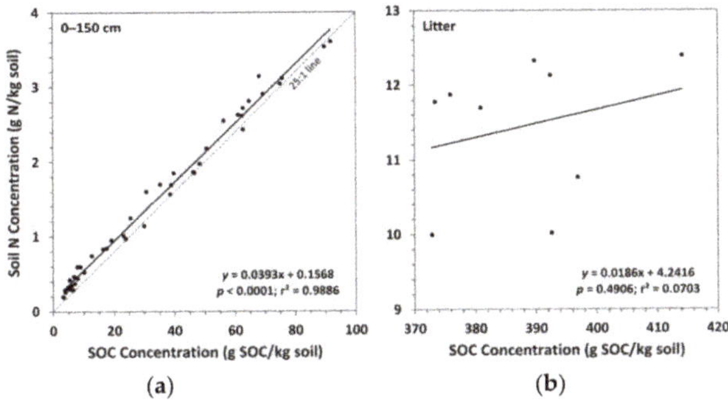

Figure 5. Soil nitrogen (N) concentration versus soil organic carbon (SOC) concentration for: (a) the whole mineral soil ($n = 45$); (b) the litter layer ($n = 9$). Dashed 25:1 line added to provide a reference for the C:N ratio for the whole mineral soil.

Soil was strongly acid across all treatments. Normalized by depth, mean soil pH values for the mineral soil for T_{trt}, FT_{trt}, and C_{trt} were 5.05, 4.98, and 5.01, respectively. In the litter layer and upper 100 cm of mineral soil, FT_{trt} had consistently lower pH values than the other two treatments (Figure 6). However, this difference was significant only in the 0–10 cm depth layer between FT_{trt} and T_{trt}.

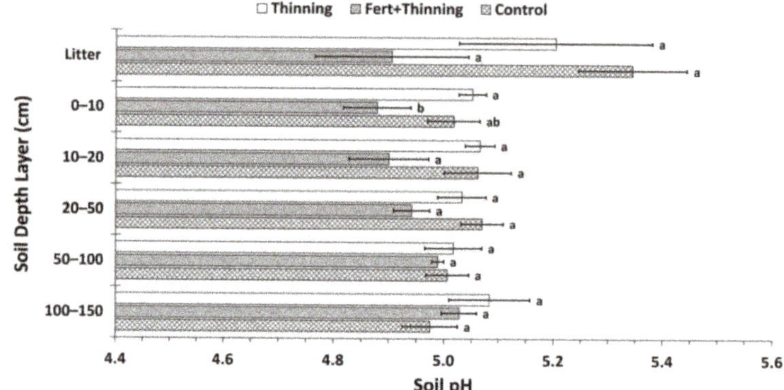

Figure 6. Mean soil pH by treatment. Sample size for all treatments and soil depth layers is three. Error bars represent ± one standard error. Means within each soil depth layer accompanied by the same letter (i.e., "a" or "b") are not significantly different (Tukey's HSD, $\alpha = 0.1$). Fert, fertilized.

3.3. Soil Organic Carbon and Nitrogen Stocks

Calculated using the fixed-depth approach, SOC and N stocks in the mineral soil were consistently highest for C_{trt} at all depths (Figure 7). Compared to T_{trt}, these differences were significant above 50 cm for SOC stocks and at all depths except the 50–100 cm soil depth layer for soil N stocks. These differences were significant in the 10–20 and 100–150 cm soil depth layers for both SOC and N stocks compared to FT_{trt}. Across all treatments, the majority of SOC and N stocks were below 20 cm ($\geq$55% and >60%, respectively) to a depth of 150 cm. FT_{trt} had a significantly greater litter layer mean SOC stock compared to C_{trt} and a significantly greater litter layer mean N stock compared to both C_{trt} and T_{trt}. Mean litter layer thickness was similar for FT_{trt} and T_{trt} (4.0 ± 0.3 and 3.8 ± 0.1 cm, respectively) and was considerably (though not significantly) thicker for these two treatments than for C_{trt} (3.1 ± 0.6 cm). Cumulative SOC stocks were significantly less at all depths below the litter layer for T_{trt} compared to C_{trt}, with a difference of 28% to a depth of 150 cm (Figure 7). The portion of this difference that occurred below 20 cm was 72%. FT_{trt} SOC stock to 150 cm in depth approximated the average of the other two treatments and was not significantly different from either. Cumulative N stocks followed similar trends (Figure 7). T_{trt} had 29% less soil N than C_{trt} to a depth of 150 cm, with 76% of this difference occurring below 20 cm.

The mass-based estimation of SOC and N stocks yielded similar results to the fixed-depth approach (Figure 8). Total differences in SOC and N stocks between C_{trt} and T_{trt} were 25% and 27%, respectively, slightly less than for the fixed-depth approach. The portion of these differences that occurred in deeper soil below ~20 cm was 56% and 64%, respectively.

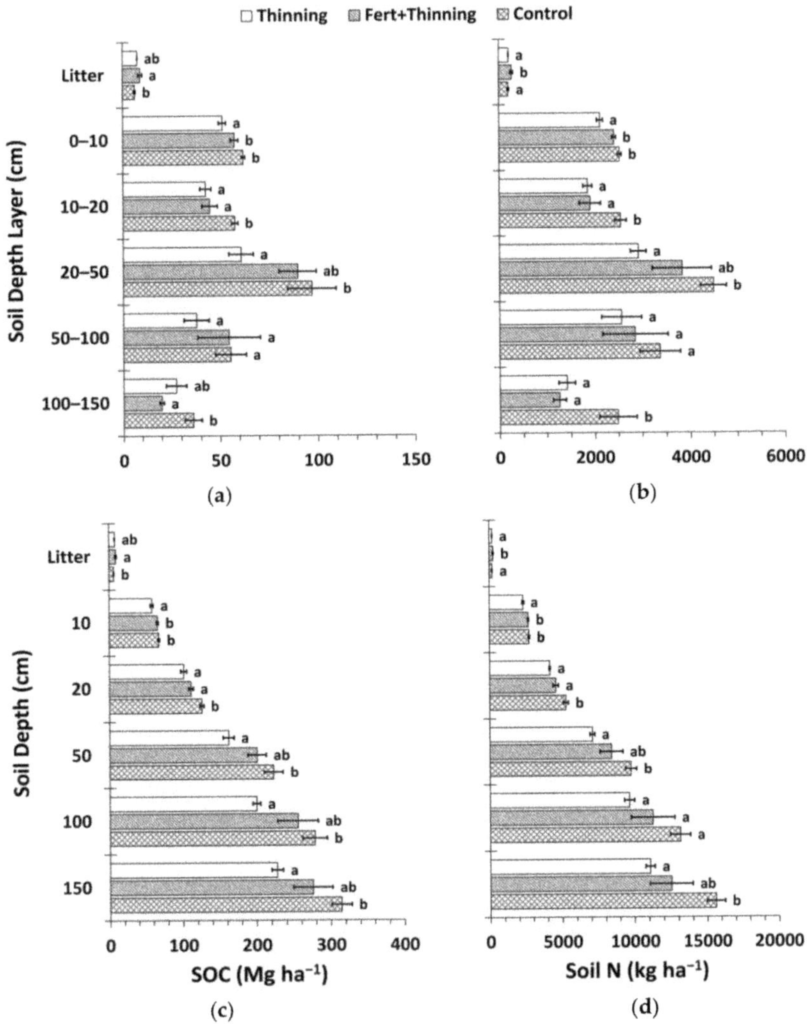

Figure 7. Mean soil organic carbon (SOC) and nitrogen (N) stocks by treatment calculated using the fixed-depth approach (Equations (2) and (3)). (**a**) Mean SOC stock by soil depth layer; (**b**) mean soil N stock by soil depth layer; (**c**) cumulative mean SOC stock; (**d**) cumulative mean soil N stock. Sample size for all treatments and soil depth layers is three. Error bars represent ± one standard error. Means within each soil depth layer accompanied by the same letter (i.e., "a" or "b") are not significantly different (Tukey's HSD, $\alpha = 0.1$). Fert, fertilized.

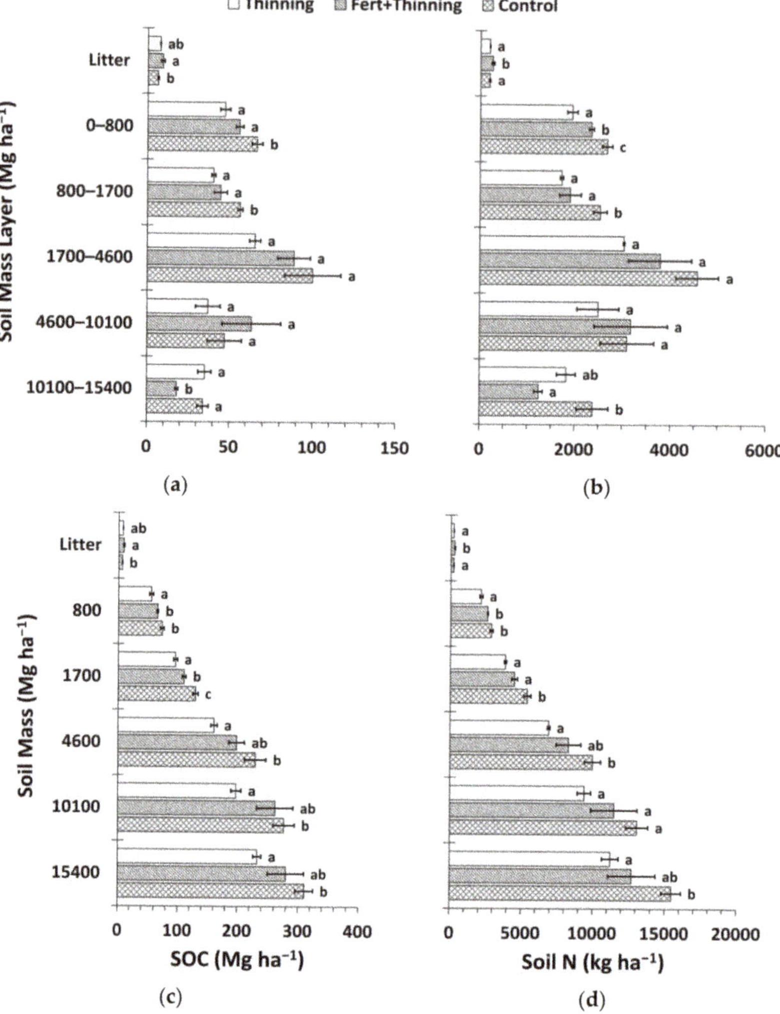

Figure 8. Mean soil organic carbon (SOC) and nitrogen (N) stocks by treatment calculated using the mass-based approach (Equation (4) and web-accessible spreadsheet [59] created by Wendt and Hauser [58]). (**a**) Mean SOC stock by soil depth layer; (**b**) mean soil N stock by soil depth layer; (**c**) cumulative mean SOC stock; (**d**) cumulative mean soil N stock. Sample size for all treatments and soil depth layers is three. Error bars represent ± one standard error. Means within each soil depth layer accompanied by the same letter (i.e., "a" or "b") are not significantly different (Tukey's HSD, $\alpha = 0.1$). Fert, fertilized.

3.4. Carbon Stocks and Sequestration

Total and belowground C stocks followed the same pattern as SOC stocks, with C_{trt} and T_{trt} having the highest and lowest C stocks, respectively (Figure 9). Trends in aboveground and root C sequestration were the opposite, with C_{trt} and T_{trt} having the lowest and highest C stocks, respectively. However, none of these differences were significant. Approximately half or more of total C stocks were contained in the soil across treatments (47%, 54%, and 58% for T_{trt}, FT_{trt}, and C_{trt}, respectively).

Initial trees per hectare was a significant predictor of aboveground ($p = 0.0031$; $r^2 = 0.7363$), root ($p = 0.0026$; $r^2 = 0.7477$), and total C stocks ($p = 0.0479$; $r^2 = 0.4503$) and was positively related to these three variables. There was no relationship between initial trees per hectare and SOC stock ($p = 0.5513$; $r^2 = 0.053$). Soil organic C stock was a significant predictor of total C stock, the two variables being positively related, and explained more of the variation in total C stock than initial trees per hectare ($p = 0.0041$; $r^2 = 0.7143$).

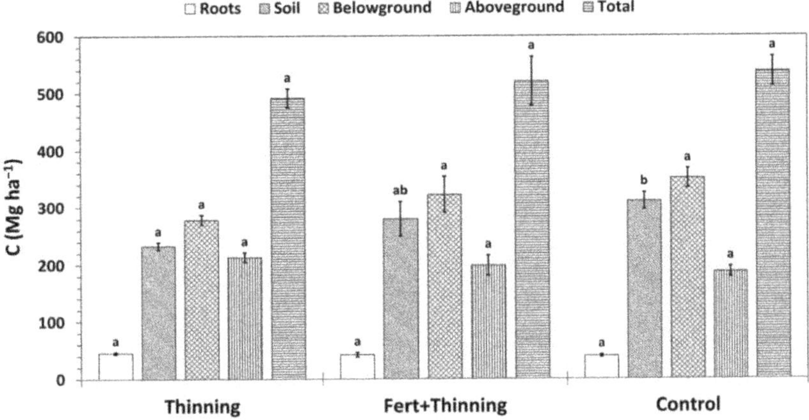

Figure 9. Mean carbon (C) stocks by treatment. Total C stocks are the sum of above- and belowground C stocks, where aboveground and root C stocks include the biomass of thinned trees, and belowground C stocks are the sum of root C and soil organic C (calculated using the mass-based approach). Sample size for all treatment groups and categories is three. Error bars represent ± one standard error. Means within each category accompanied by the same letter (i.e., "a" or "b") are not significantly different (Tukey's HSD, $\alpha = 0.1$). Fert, fertilized.

3.5. Merchantable Volume

Merchantable volume, calculated in thousands of board-feet per acre (as 9.8-m long logs to a 15-cm top diameter), was significantly greater for FT_{trt} than T_{trt}, a difference of ~11% (41.1 ± 0.4, 45.6 ± 1.3, and 43.0 ± 1.2 for T_{trt}, FT_{trt}, and C_{trt}, respectively). No other variable (e.g., SOC or N stock, initial trees per hectare, and DBH) was significantly related to merchantable volume. Final mean DBH (cm) was greatest for FT_{trt} and lowest for T_{trt}, but these differences were not significant (13.0 ± 0.9, 16.2 ± 1.4, and 14.0 ± 1.6 for T_{trt}, FT_{trt}, and C_{trt}, respectively). Initial trees per hectare was a significant predictor of DBH, the two variables being negatively related ($p = 0.0024$; $r^2 = 0.7529$).

4. Discussion

It is important to first determine which approach, fixed-depth or mass-based, provided a more accurate quantification and comparison of SOC and N stocks. While the fixed-depth and mass-based approaches resulted in similar conclusions overall, the mass-based approach appears to have better represented the degree of change along the vertical soil profile by comparing equal soil masses and eliminating bulk density as a factor. Because bulk density was significantly higher for T_{trt} than C_{trt} above 20 cm, the mass-based approach resulted in greater SOC and N stock differences between these two treatments in the surface soil layers compared to the fixed-depth approach. Significant differences in bulk density observed between treatments were most likely the result of changes in SOC concentrations. This conclusion is supported by the lack of heavy equipment used during thinning treatments (i.e., a lack of soil-compacting operations), as well as the fact that significant differences in bulk density between treatments coincided with significant differences in SOC concentrations and occurred in soil layers where SOC concentration was a significant predictor of bulk density. In order to account and correct for changes in bulk density with time (that have not resulted from soil erosion or deposition), the mass-based approach is increasingly recommended for SOC and other soil-nutrient inventories [58,60,70–73]. Soil organic C and N stocks quantified using the mass-based approach and compared among equal soil masses will be considered the more accurate account in the current study, as this process removed changes in bulk density as a confounding variable. Further discussion will refer to the mass-based approach and mass-based calculations of SOC and N stocks.

Litter layer SOC and N stock differences between FT_{trt} and C_{trt} primarily resulted from the greater thickness of FT_{trt} litter layer compared to C_{trt}, rather than from differences in SOC and N concentrations. This greater litter layer thickness may have been due to increased N availability post-fertilization treatments and thus increased understory biomass and turnover [12]. The significantly lower litter layer soil N concentration of T_{trt} compared to the other two treatments likely resulted from forest floor organic matter additions with high C:N ratios (such as coarse woody debris left as slash on the forest floor) and subsequent microbial N scavenging in this layer [21,22]. High microbial demand for N during organic-matter decomposition for FT_{trt} post-thinning was likely compensated for by the addition of N via fertilization treatments.

In the mineral soil, SOC and N stock differences between the treatments were due to differences in SOC and N concentrations. Cumulatively, T_{trt} and FT_{trt} contained 77.6 and 30.5 Mg ha^{-1} less SOC, respectively, than C_{trt} to a depth of ~150 cm. The difference between T_{trt} and C_{trt} SOC stocks was significant and occurred over a shorter post-treatment timeframe (~11 years) than the difference observed between FT_{trt} and C_{trt}. Assuming equivalent SOC stocks pre-treatment, the rate of post-treatment SOC loss for T_{trt} was ~700 g C m^{-2} y^{-1} compared to C_{trt}. The typical range of SOC accumulation rates in temperate forest soils is ~2 to 70 g C m^{-2} y^{-1} [74], with an average rate of 34 g C m^{-2} y^{-1}. Raich and Schlesinger [75] reported a mean soil respiration rate for temperate coniferous forests of 681 g C m^{-2} y^{-1}. However, rates ranging between ~950 and 1750 g C m^{-2} y^{-1} have been measured [19,20,76]. Using average C flux rates, we estimate a potential T_{trt} post-treatment respiration rate of ~1350 g C m^{-2} y^{-1}, which falls within the published range of soil respiration rates for temperate coniferous forests. Of course, other mechanisms of SOC (and N) loss, such as increased leaching and export, could also help explain observed differences in SOC and N stocks between treatments.

Decreased radiation interception by trees post-thinning could have increased soil temperatures, enhancing microbial metabolic activity and potentially accounting for some of the loss of SOC and N from T_{trt} compared to C_{trt} [16–20]. Additionally, greater bulk density in the surface soil layers of T_{trt} compared to C_{trt} could have increased heat-transfer rates to deeper soil layers [20,21] where the majority of the SOC stock was contained and also lost. Fresh C inputs and the creation of preferential flow paths due to the mineralization of root biomass, which can persist for many years following the harvest of trees [31], and increases in DOC flux due to additional organic matter inputs and decreased transpiration and rain interception [29,30] could have increased SOC decomposition via priming post-thinning [23–25,32–34]. The soil sampled in the current study may be particularly vulnerable to rapid SOC decomposition when subjected to changing environmental conditions and potential priming effects, as it is an older soil that likely developed and accrued SOC over hundreds of thousands to millions of years under relatively stable conditions [19,23–28,77].

In the surface soil layers (0–20 cm in depth), which contain the majority of root [78] and microbial [79] biomass, the mean C:N ratio across treatments was ~24, suggesting tight N cycling and the potential for N scavenging, particularly after the addition of high C:N slash to the forest floor post-thinning [21,22]. However, nitrate leaching, which has been observed to increase under slash left on the forest floor [35–39], is also a possible mechanism of N loss and SOC loss via priming [24,25]. FT_{trt} may have been particularly vulnerable to nitrate leaching post-thinning treatments due to reduced N uptake by roots coinciding with N fertilization. The lower pH in the upper 100 cm of mineral soil and litter layer of FT_{trt} compared to the other two treatments indicates that a considerable portion of nitrate resulting from the nitrification of urea potentially was not taken up by plants, causing the net addition of one proton (H^+) to the soil solution per urea compound [80]. Nitrate leaching would have been generally promoted by the high precipitation in the region studied. Notably, several of the excavated pits had redoximorphic features as high as 75 cm in depth, indicating a relatively high or perched water table. A previous study conducted at the same site identified soil features (massive, clay cemented) below 3 m that could result in a perched water table [81]. As Douglas-fir roots commonly extend to at least 3 m in depth [47], thinning treatments may have sufficiently reduced transpiration to allow a local rise in groundwater level, particularly during the wet season. This rise would have enabled the groundwater to contact more superficial soil layers that have higher DOC and N concentrations, potentially increasing the export of DOC and various forms of N [37,41].

Although SOC stocks were significantly greater for C_{trt} than T_{trt}, differences in belowground and total C stocks between these two treatments were diluted due to T_{trt} having greater aboveground and root C stocks, which increased with initial trees per hectare. On average, T_{trt} had substantially higher initial trees per hectare than C_{trt} (952 and 682 trees ha^{-1}, respectively). Interestingly, there was no relationship between SOC stock and initial trees per hectare. This is somewhat counterintuitive when thinning treatments, which did affect SOC stocks, were carried out similarly within about a decade of the reductions in initial trees per hectare. However, several key differences may explain this phenomenon. While thinning treatments occurred after crown closure, thus exposing previously shaded and covered areas, reductions in initial trees per hectare occurred prior to crown closure and would not have drastically changed soil conditions. More substantial understory cover prior to crown closure likely insulated the soil from significant temperature changes and increased N uptake following reductions in initial trees per hectare, decreasing nitrate leaching and helping to retain N on-site [82,83]. At this early stage in stand and plant development, leaves and roots would have had a lower C:N ratio [84], reducing the potential for N scavenging during the decomposition of organic matter left on-site. Additionally, roots would have been less dense and rooted less deeply in the soil, likely decreasing DOC flux and the potential for priming effects compared to thinning treatments implemented years later. Differences in soil microbial communities can also lead to differences in SOC and nutrient dynamics. Smith et al. [85] found that soil microbial communities differed between younger and older forests in a study examining forests aged 20 years and older. Although initial trees per hectare were reduced at the site in the current study when the stand was aged <15 years, differences in ground and soil conditions—and thus differences in microbial communities—would likely be greater between juvenile and adult stands than between younger and older adult stands. The lack of relationship between SOC stock and initial trees per hectare suggests that potentially lower SOC accumulation over time due to reduced root C inputs (from reduced root biomass) post-thinning did not play a substantial role in decreasing the SOC stock of T_{trt} compared to C_{trt}.

Despite greater merchantable volume for FT_{trt} compared to C_{trt}, any net monetary gains would have been minimal due to the additional expenses of fertilization and thinning treatments. T_{trt} resulted in the least financial gain over the length of the rotation (i.e., it had the lowest merchantable volume in addition to incurring thinning expenses) and reduced soil quality and nutrient stocks for the succeeding rotation. On the other hand, reducing the initial trees per hectare prior to crown closure did not affect SOC and N stocks and the variation in total C stock was explained more by SOC stock (71%) than by initial trees per hectare (45%). Therefore, the typically low-cost practice of reducing initial trees per hectare could provide benefits such as increased stand stability, health, and DBH, while potentially avoiding negative effects such as reducing SOC and N stocks.

5. Conclusions

This long-term study shows that forest-management practices can affect both surface and deep SOC and N stocks on decadal timescales. Thinning treatments reduced SOC and N stocks by 25% and 27%, respectively, with most of this loss occurring below ~20 cm to a depth of ~150 cm. Changing the soil environment by affecting the ecosystem, C inputs, roots, and soil properties can increase

SOC decomposition, N mineralization, and SOC and N export. In this study, a combination of these factors and their complex interactions likely resulted in the observed decreases in SOC and N stocks post-thinning. Although thinning treatments had a negative effect on SOC and N stocks, reducing the initial trees per hectare prior to crown closure is a low-cost alternative practice that could provide the stand benefits associated with thinning a more mature stand without decreasing soil fertility and, potentially, the productivity of future stands. Additional field studies should examine the mechanisms behind SOC and N losses post-thinning (particularly in deeper soil layers) and the effects on stand and soil dynamics of reducing the initial trees per hectare at various intensities and stages prior to crown closure. As the majority of SOC and N stocks are contained in deeper soil layers, accurately assessing SOC and N budgets and comparing changes over time requires sampling soil deeper than 20 cm.

Author Contributions: C.D.G., R.B.H., and E.C.T. conceived and designed the experiment; C.D.G. collected and analyzed the samples with assistance and feedback from J.N.J.; C.D.G. analyzed the data with feedback from E.C.T., J.N.J., and R.B.H.; R.B.H. contributed reagents/materials/analysis tools; C.D.G. wrote the paper.

Funding: This research was funded by the Stand Management Cooperative, School of Environmental and Forest Sciences, University of Washington, Seattle, Washington.

Acknowledgments: We thank the members and staff of the University of Washington Stand Management Cooperative for funding this research and providing valuable feedback, as well as for establishing and maintaining the study site and for providing technical support and assistance. Thanks also go to Dongsen Xue, Hanzhang Ding "Chris," and Tony Scigliano for assistance in the laboratory, and to Patrick Tobin and Darlene Zabowski for their feedback. Finally, we thank the three anonymous reviewers whose constructive comments helped improve the quality of this paper.

Conflicts of Interest: The authors declare no conflict of interest.

References

1. Bellassen, V.; Luyssaert, S. Carbon sequestration: Managing forests in uncertain times. *Nature* **2014**, *506*, 153–155. [CrossRef] [PubMed]
2. Pan, Y.; Birdsey, R.A.; Fang, J.; Houghton, R.; Kauppi, P.E.; Kurz, W.A.; Phillips, O.L.; Shvidenko, A.; Lewis, S.L.; Canadell, J.G.; et al. A large and persistent carbon sink in the world's forests. *Science* **2011**, *333*, 988–993. [CrossRef] [PubMed]
3. Lal, R. Forest soils and carbon sequestration. *For. Ecol. Manag.* **2005**, *220*, 242–258. [CrossRef]
4. Stockmann, U.; Adams, M.A.; Crawford, J.W.; Field, D.J.; Henakaarchchi, N.; Jenkins, M.; Minasny, B.; McBratney, A.B.; de Courcelles, V.D.R.; Singh, K.; et al. The knowns, known unknowns and unknowns of sequestration of soil organic carbon. *Agric. Ecosyst. Environ.* **2013**, *164*, 80–99. [CrossRef]
5. Clarke, N.; Gundersen, P.; Jönsson-Belyazid, U.; Kjønaas, O.J.; Persson, T.; Sigurdsson, B.D.; Stupak, I.; Vesterdal, L. Influence of different tree-harvesting intensities on forest soil carbon stocks in boreal and northern temperate forest ecosystems. *For. Ecol. Manag.* **2015**, *351*, 9–19. [CrossRef]
6. Jobbágy, E.G.; Jackson, R.B. The vertical distribution of soil organic carbon and its relation to climate and vegetation. *Ecol. Appl.* **2000**, *10*, 423–436. [CrossRef]
7. Dixon, R.K.; Brown, S.; Houghton, R.A.; Solomon, A.M.; Trexler, M.C.; Wisniewski, J. Carbon pools and flux of global forest ecosystems. *Science* **1994**, *263*, 185–190. [CrossRef] [PubMed]
8. Milne, E.; Banwart, S.A.; Noellemeyer, E.; Abson, D.J.; Ballabio, C.; Bampa, F.; Bationo, A.; Batjes, N.H.; Bernoux, M.; Bhattacharyya, T.; et al. Soil carbon, multiple benefits. *Environ. Dev.* **2015**, *13*, 33–38. [CrossRef]
9. Carter, R.E.; Klinka, K. Relationships between growing-season soil water-deficit, mineralizable soil nitrogen and site index of coastal Douglas fir. *For. Ecol. Manag.* **1990**, *30*, 301–311. [CrossRef]
10. Blake, J.I.; Chappell, H.N.; Bennett, W.S.; Gessel, S.P.; Webster, S.R. Douglas fir growth and foliar nutrient responses to nitrogen and sulfur fertilization. *Soil Sci. Soc. Am. J.* **1990**, *54*, 257. [CrossRef]
11. Chappell, H.N.; Cole, D.W.; Gessel, S.P.; Walker, R.B. Forest fertilization research and practice in the Pacific Northwest. *Fertil. Res.* **1991**, *27*, 129–140. [CrossRef]
12. Jandl, R.; Lindner, M.; Vesterdal, L.; Bauwens, B.; Baritz, R.; Hagedorn, F.; Johnson, D.W.; Minkkinen, K.; Byrne, K.A. How strongly can forest management influence soil carbon sequestration? *Geoderma* **2007**, *137*, 253–268. [CrossRef]

13. Li, W.; Jin, C.; Guan, D.; Wang, Q.; Wang, A.; Yuan, F.; Wu, J. The effects of simulated nitrogen deposition on plant root traits: A meta-analysis. *Soil Biol. Biochem.* **2015**, *82*, 112–118. [CrossRef]

14. James, J.; Harrison, R. The effect of harvest on forest soil carbon: A meta-analysis. *Forests* **2016**, *7*, 308. [CrossRef]

15. Bird, J.A.; Kleber, M.; Torn, M.S. ^{13}C and ^{15}N stabilization dynamics in soil organic matter fractions during needle and fine root decomposition. *Org. Geochem.* **2008**, *39*, 465–477. [CrossRef]

16. Chase, C.W.; Kimsey, M.J.; Shaw, T.M.; Coleman, M.D. The response of light, water, and nutrient availability to pre-commercial thinning in dry inland Douglas-fir forests. *For. Ecol. Manag.* **2016**, *363*, 98–109. [CrossRef]

17. Cheng, X.; Han, H.; Kang, F.; Liu, K.; Song, Y.; Zhou, B.; Li, Y. Short-term effects of thinning on soil respiration in a pine (*Pinus tabulaeformis*) plantation. *Biol. Fertil. Soils* **2014**, *50*, 357–367. [CrossRef]

18. Thibodeau, L.; Raymond, P.; Camiré, C.; Munson, A.D. Impact of precommercial thinning in balsam fir stands on soil nitrogen dynamics, microbial biomass, decomposition, and foliar nutrition. *Can. J. For. Res.* **2000**, *30*, 229–238. [CrossRef]

19. Hicks Pries, C.E.; Castanha, C.; Porras, R.; Torn, M.S. The whole-soil carbon flux in response to warming. *Science* **2017**, *355*, 1420–1423. [CrossRef] [PubMed]

20. Gallo, A.C. Response of Soil Temperature, Moisture, and Respiration Two Years Following Intensive Organic Matter and Compaction Manipulations in Oregon Cascade Forests. MS Thesis, Oregon State University, Corvallis, OR, USA, 2016.

21. Brady, N.C.; Weil, R.R. *The Nature and Properties of Soils*, 14th ed.; Pearson: Upper Saddle River, NJ, USA, 2008.

22. Staaf, H.; Berg, B. Accumulation and release of plant nutrients in decomposing Scots pine needle litter. Long-term decomposition in a Scots pine forest II. *Can. J. Bot.* **1982**, *60*, 1561–1568. [CrossRef]

23. Fontaine, S.; Barot, S.; Barré, P.; Bdioui, N.; Mary, B.; Rumpel, C. Stability of organic carbon in deep soil layers controlled by fresh carbon supply. *Nature* **2007**, *450*, 277–280. [CrossRef] [PubMed]

24. Carrillo, Y.; Bell, C.; Koyama, A.; Canarini, A.; Boot, C.M.; Wallenstein, M.; Pendall, E. Plant traits, stoichiometry and microbes as drivers of decomposition in the rhizosphere in a temperate grassland. *J. Ecol.* **2017**, *105*, 1750–1765. [CrossRef]

25. Kalbitz, K.; Solinger, S.; Park, J.-H.; Michalzik, B.; Matzner, E. Controls on the dynamics of dissolved organic matter in soils: A review. *Soil Sci.* **2000**, *165*, 277–304. [CrossRef]

26. Schmidt, M.W.I.; Torn, M.S.; Abiven, S.; Dittmar, T.; Guggenberger, G.; Janssens, I.A.; Kleber, M.; Kögel-Knabner, I.; Lehmann, J.; Manning, D.A.; et al. Persistence of soil organic matter as an ecosystem property. *Nature* **2011**, *478*, 49–56. [CrossRef] [PubMed]

27. Harrison, R.B.; Footen, P.W.; Strahm, B.D. Deep soil horizons: Contribution and importance to soil carbon pools and in assessing whole-ecosystem response to management and global change. *For. Sci.* **2011**, *57*, 67–76.

28. Hicks Pries, C.E.; Schuur, E.A.G.; Natali, S.M.; Crummer, K.G. Old soil carbon losses increase with ecosystem respiration in experimentally thawed tundra. *Nat. Clim. Chang.* **2016**, *6*, 214–218. [CrossRef]

29. Neff, J.C.; Asner, G.P. Dissolved organic carbon in terrestrial ecosystems: Synthesis and a model. *Ecosystems* **2001**, *4*, 29–48. [CrossRef]

30. Sanderman, J.; Baldock, J.A.; Amundson, R.; Baldock, J.A. Dissolved organic carbon chemistry and dynamics in contrasting forest and grassland soils. *Biogeochemistry* **2008**, *89*, 181–198. [CrossRef]

31. Strahm, B.D.; Harrison, R.B.; Terry, T.A.; Harrington, T.B.; Adams, A.B.; Footen, P.W. Changes in dissolved organic matter with depth suggest the potential for postharvest organic matter retention to increase subsurface soil carbon pools. *For. Ecol. Manag.* **2009**, *258*, 2347–2352. [CrossRef]

32. Deb, S.K.; Shukla, M.K. A review of dissolved organic matter transport processes affecting soil and environmental quality. *J. Environ. Anal. Toxicol.* **2011**, *1*. [CrossRef]

33. Bundt, M.; Widmer, F.; Pesaro, M.; Zeyer, J.; Blaser, P. Preferential flow paths: Biological "hot spots" in soils. *Soil Biol. Biochem.* **2001**, *33*, 729–738. [CrossRef]

34. Hagedorn, F.; Bundt, M. The age of preferential flow paths. *Geoderma* **2002**, *108*, 119–132. [CrossRef]

35. Strahm, B.D.; Harrison, R.B.; Terry, T.A.; Flaming, B.L.; Licata, C.W.; Petersen, K.S. Soil solution nitrogen concentrations and leaching rates as influenced by organic matter retention on a highly productive Douglas-fir site. *For. Ecol. Manag.* **2005**, *218*, 74–88. [CrossRef]

36. Emmett, B.A.; Anderson, J.M.; Hornung, M. The controls on dissolved nitrogen losses following two intensities of harvesting in a Sitka spruce forest (N. Wales). *For. Ecol. Manag.* **1991**, *41*, 65–80. [CrossRef]

37. Nieminen, M. Export of dissolved organic carbon, nitrogen and phosphorus following clear-cutting of three Norway spruce forests growing on drained peatlands in southern Finland. *Silva Fenn.* **2004**, *38*, 123–132. [CrossRef]

38. Staaf, H.; Olsson, B.A. Effects of slash removal and stump harvesting on soil water chemistry in a clearcutting in SW Sweden. *Scand. J. For. Res.* **1994**, *9*, 305–310. [CrossRef]

39. Rosén, K.; Lundmark-Thelin, A. Increased nitrogen leaching under piles of slash—A consequence of modern forest harvesting techniques. *Scand. J. For. Res.* **1987**, *2*, 21–29. [CrossRef]

40. Fox, T.R. Nitrogen mineralization following fertilization of Douglas-fir forests with urea in western Washington. *Soil Sci. Soc. Am. J.* **2004**, *68*, 1720–1728. [CrossRef]

41. Laudon, H.; Hedtjärn, J.; Schelker, J.; Bishop, K.; Sørensen, R.; Agren, A. Response of dissolved organic carbon following forest harvesting in a boreal forest. *Ambio* **2009**, *38*, 381–386. [CrossRef] [PubMed]

42. Mobley, M.L.; Lajtha, K.; Kramer, M.G.; Bacon, A.R.; Heine, P.R.; Richter, D.D. Surficial gains and subsoil losses of soil carbon and nitrogen during secondary forest development. *Glob. Chang. Biol.* **2015**, *21*, 986–996. [CrossRef] [PubMed]

43. Diochon, A.C.; Kellman, L. Physical fractionation of soil organic matter: Destabilization of deep soil carbon following harvesting of a temperate coniferous forest. *J. Geophys. Res. Biogeosci.* **2009**, *114*, 1–9. [CrossRef]

44. Richter, D.D.; Markewitz, D.; Trumbore, S.E.; Wells, C.G. Rapid accumulation and turnover of soil carbon in a re-establishing forest. *Nature* **1999**, *400*, 56–58. [CrossRef]

45. Batjes, N.H. Total carbon and nitrogen in the soils of the world. *Eur. J. Soil Sci.* **1996**, *47*, 151–163. [CrossRef]

46. Tarnocai, C.; Canadell, J.G.; Schuur, E.A.G.; Kuhry, P.; Mazhitova, G.; Zimov, S. Soil organic carbon pools in the northern circumpolar permafrost region. *Glob. Biogeochem. Cycles* **2009**, *23*. [CrossRef]

47. Stone, E.L.; Kalisz, P.J. On the maximum extent of tree roots. *For. Ecol. Manag.* **1991**, *46*, 59–102. [CrossRef]

48. Jandl, R.; Rodeghiero, M.; Martinez, C.; Cotrufo, M.F.; Bampa, F.; van Wesemael, B.; Harrison, R.B.; Guerrini, I.A.; deB Richter, D., Jr.; Rustad, L.; et al. Current status, uncertainty and future needs in soil organic carbon monitoring. *Sci. Total Environ.* **2014**, *468–469*, 376–383. [CrossRef] [PubMed]

49. Curtis, R.O. A simple index of stand density for Douglas-fir. *For. Sci.* **1982**, *28*, 92–94.

50. PRISM Climate Group. Oregon State University, Corvallis, OR, USA, created 4 Feb 2004. Available online: http://prism.oregonstate.edu (accessed on 17 February 2017).

51. Gross, C.D.; Harrison, R.B. Quantifying and comparing soil carbon stocks: Underestimation with the core sampling method. *Soil Sci. Soc. Am. J.* **2018**, doi:10.2136/sssaj2018.01.0015.

52. Hamburg, S.P. Simple rules for measuring changes in ecosystem carbon in forestry-offset projects. *Mitag. Adapt. Strateg. Glob. Chang.* **2000**, *5*, 25–37. [CrossRef]

53. Holub, S.M. *Soil Carbon Change in Pacific Northwest Coastal Douglas-Fir Forests: Change Detection Following Harvest—Soils Establishment Report*; Weyerhaeuser NR, Timberlands Technology, Production Forestry West: Albany, OR, USA, 2011.

54. Harrison, R.B.; Adams, A.B.; Licata, C.; Flaming, B.; Wagoner, G.L.; Carpenter, P.; Vance, E.D. Quantifying deep-soil and coarse-soil fractions: Avoiding sampling bias. *Soil Sci. Soc. Am. J.* **2003**, *67*, 1602–1606. [CrossRef]

55. Soil Survey Staff, Natural Resources Conservation Service, United States Department of Agriculture. Web Soil Survey. Available online: https://websoilsurvey.sc.egov.usda.gov/ (accessed on 5 July 2017).

56. Walthert, L.; Graf, U.; Kammer, A.; Luster, J.; Pezzotta, D.; Zimmermann, S.; Hagedorn, F. Determination of organic and inorganic carbon, δ13C, and nitrogen in soils containing carbonates after acid fumigation with HCl. *J. Plant Nutr. Soil Sci.* **2010**, *173*, 207–216. [CrossRef]

57. Tisdall, A.L. Comparison of methods of determining apparent density of soils. *Aust. J. Agric. Res.* **1951**, *2*, 349–354. [CrossRef]

58. Wendt, J.W.; Hauser, S. An equivalent soil mass procedure for monitoring soil organic carbon in multiple soil layers. *Eur. J. Soil Sci.* **2013**, *64*, 58–65. [CrossRef]

59. Wendt, J.W. ESM Sample Spreadsheets, 2012. Available online: https://drive.google.com/file/d/0BzxNFfzLbFxjSG9RWlpwQ0FXc0k/view?usp=sharing (accessed on 16 May 2017).

60. Lee, J.; Hopmans, J.W.; Rolston, D.E.; Baer, S.G.; Six, J. Determining soil carbon stock changes: Simple bulk density corrections fail. *Agric. Ecosyst. Environ.* **2009**, *134*, 251–256. [CrossRef]

61. Throop, H.L.; Archer, S.R.; Monger, H.C.; Waltman, S. When bulk density methods matter: Implications for estimating soil organic carbon pools in rocky soils. *J. Arid Environ.* **2012**, *77*, 66–71. [CrossRef]

62. Zabowski, D.; Whitney, N.; Gurung, J.; Hatten, J. Total soil carbon in the coarse fraction and at depth. *For. Sci.* **2011**, *57*, 11–18. [CrossRef]

63. Corti, G.; Ugolini, F.C.; Agnelli, A.; Certini, G.; Cuniglio, R.; Berna, F.; Fernández Sanjurjo, M.J. The soil skeleton, a forgotten pool of carbon and nitrogen in soil. *Eur. J. Soil Sci.* **2002**, *53*, 283–298. [CrossRef]

64. Whitney, N.; Zabowski, D. Soil total nitrogen in the coarse fraction and at depth. *Soil Sci. Soc. Am. J.* **2004**, *68*, 612–619. [CrossRef]

65. Jenkins, J.C.; Chojnacky, D.C.; Heath, L.S.; Birdsey, R.A. National-scale biomass estimators for United States tree species. *For. Sci.* **2003**, *49*, 12–35.

66. Miles, P.D.; Smith, W.B. *Specific Gravity and Other Properties of Wood and Bark for 156 Tree Species Found in North America*; US Department of Agriculture, Forest Service, Northern Research Station: Newtown Square, PA, USA, 2009.

67. Birdsey, R.A. *Carbon Storage and Accumulation in United States Forest Ecosystems*; United States Department of Agriculture, Forest Service, Northeastern Forest Experiment Station: Radnor, PA, USA, 1992.

68. Zar, J.H. *Biostatistical Analysis*, 5th ed.; Prentice-Hall/Pearson: Upper Saddle River, NJ, USA, 2010.

69. RStudio Team. *RStudio: Integrated Development for R*; RStudio Inc.: Boston, MA, USA, 2016.

70. Schrumpf, M.; Schulze, E.D.; Kaiser, K.; Schumacher, J. How accurately can soil organic carbon stocks and stock changes be quantified by soil inventories? *Biogeosciences* **2011**, *8*, 1193–1212. [CrossRef]

71. Wuest, S.B. Correction of bulk density and sampling method biases using soil mass per unit area. *Soil Sci. Soc. Am. J.* **2009**, *73*, 312. [CrossRef]

72. Mikha, M.M.; Benjamin, J.G.; Halvorson, A.D.; Nielsen, D.C. Soil carbon changes influenced by soil management and calculation method. *Open J. Soil Sci.* **2013**, *3*, 123–131. [CrossRef]

73. Ellert, B.H.; Bettany, J.R. Calculation of organic matter and nutrients stored in soils under contrasting management regimes. *Can. J. Soil Sci.* **1995**, *75*, 529–538. [CrossRef]

74. Post, W.M.; Kwon, K.C.A. Soil carbon sequestration and land-use change: Processes and potential. *Glob. Chang. Biol.* **2000**, *6*, 317–327. [CrossRef]

75. Raich, J.W.; Schlesinger, W.H. The global carbon dioxide flux in soil respiration and its relationship to vegetation and climate. *Tellus B* **1992**, *44*, 81–99. [CrossRef]

76. Tang, J.; Qi, Y.; Xu, M.; Misson, L.; Goldstein, A.H. Forest thinning and soil respiration in a ponderosa pine plantation in the Sierra Nevada. *Tree Physiol.* **2005**, *25*, 57–66. [CrossRef] [PubMed]

77. Lin, H. Three principles of soil change and pedogenesis in time and space. *Soil Sci. Soc. Am. J.* **2011**, *75*, 2049. [CrossRef]

78. Jackson, R.B.; Canadell, J.; Ehleringer, J.R.; Mooney, H.A.; Sala, O.E.; Schulze, E.D. A global analysis of root distributions for terrestrial biomes. *Oecologia* **1996**, *108*, 389–411. [CrossRef] [PubMed]

79. Fierer, N.; Schimel, J.P.; Holden, P.A. Variations in microbial community composition through two soil depth profiles. *Soil Biol. Biochem.* **2003**, *35*, 167–176. [CrossRef]

80. Davidson, S. Combating soil acidity: Three approaches. *Rural Res.* **1987**, *134*, 4–10.

81. James, J.N.; Dietzen, C.; Furches, J.C.; Harrison, R.B. Lessons on buried horizons and pedogenesis from deep forest soils. *Soil Horizons* **2015**, *56*. [CrossRef]

82. Chang, S.X.; Preston, C.M. Understorey competition affects tree growth and fate of fertilizer-applied ^{15}N in a coastal British Columbia plantation forest: 6-year results. *Can. J. For. Res.* **2000**, *30*, 1379–1388. [CrossRef]

83. Footen, P.W.; Harrison, R.B.; Strahm, B.D. Long-term effects of nitrogen fertilization on the productivity of subsequent stands of Douglas-fir in the Pacific Northwest. *For. Ecol. Manag.* **2009**, *258*, 2194–2198. [CrossRef]

84. Zhang, H.; Wu, H.; Yu, Q.; Wang, Z.; Wei, C.; Long, M.; Kattge, J.; Smith, M.; Han, X. Sampling date, leaf age and root size: Implications for the study of plant C:N:P stoichiometry. *PLoS ONE* **2013**, *8*. [CrossRef] [PubMed]

85. Smith, A.P.; Marín-Spiotta, E.; Balser, T. Successional and seasonal variations in soil and litter microbial community structure and function during tropical postagricultural forest regeneration: A multiyear study. *Glob. Chang. Biol.* **2015**, *21*, 3532–3547. [CrossRef] [PubMed]

 *forests*

Article

Diversity and Enzyme Activity of Ectomycorrhizal Fungal Communities Following Nitrogen Fertilization in an Urban-Adjacent Pine Plantation

Chen Ning [1,2,3,*], **Gregory M. Mueller** [2,3], **Louise M. Egerton-Warburton** [2,3], **Andrew W. Wilson** [2,3,4], **Wende Yan** [1] **and Wenhua Xiang** [1]

[1]   Faculty of Life Science and Technology, Central South University of Forestry and Technology, Changsha 410004, China; csfuywd@hotmail.com (W.Y.); xiangwh2005@163.com (W.X.)

[2]   Program in Plant Biology and Conservation, Northwestern University, Evanston, IL 60208, USA; gmueller@chicagobotanic.org (G.M.M.); LWarburton@chicagobotanic.org (L.M.E.-W.); andrew.wilson@botanicgardens.org (A.W.W.)

[3]   Chicago Botanic Garden, Glencoe, IL 60022, USA

[4]   Denver Botanic Gardens, Sam Mitchel Herbarium of Fungi, Denver, CO 80206, USA

*   Correspondence: chenning2012@u.northwestern.edu; Tel.: +1-312-404-7136

Received: 2 January 2018; Accepted: 23 February 2018; Published: 25 February 2018

Abstract: Rapid economic development and accelerated urbanization in China has resulted in widespread atmospheric nitrogen (N) deposition. One consequence of N deposition is the alteration of mycorrhizal symbioses that are critical for plant resource acquisition (nitrogen, N, phosphorus, P, water). In this study, we characterized the diversity, composition, and functioning of ectomycorrhizal (ECM) fungal communities in an urban-adjacent *Pinus elliottii* plantation under ambient N deposition (~24 kg N ha^{-1} year^{-1}), and following N fertilization (low N, 50 kg N ha^{-1} year^{-1}; high N, 300 kg N ha^{-1} year^{-1}). ECM functioning was expressed as the potential activities of extracellular enzymes required for organic N (protease), P (phosphomonoesterase), and recalcitrant polymers (phenol oxidase). Despite high ambient N deposition, ECM community composition shifted under experimental N fertilization, and those changes were linked to disparate levels of soil minerals (P, K) and organic matter (but not N), a decline in acid phosphatase (AP), and an increase in phenol oxidase (PO) potential activities. Based on enzyme stoichiometry, medium-smooth exploration type ECM species invested more in C acquisition (PO) relative to P (AP) following high N fertilization than other exploration types. ECM species with hydrophilic mantles also showed higher enzymatic PO:AP ratios than taxa with hydrophobic mantles. Our findings add to the accumulating evidence that shifts in ECM community composition and taxa specialized in organic C, N, and P degradation could modulate the soil nutrient cycling in forests exposed to chronic elevated N input.

Keywords: extracellular enzymes; hyphal exploration strategy; China; atmospheric nitrogen deposition; *Russula*

1. Introduction

Atmospheric nitrogen (N) deposition has more than doubled the inputs of N into many forest systems. One consequence of N deposition has been the increase in forest productivity [1]. Another is the alteration of soil microbial communities, and especially mycorrhizal communities [2–4]. Most forest trees, such as Pinaceae, depend on symbioses with ectomycorrhizal fungi (ECM) for resource uptake (nitrogen, N, phosphorus, P, water) from soil, and different ECM taxa appear to specialize in various forms of organic resources (reviewed in Lilleskov et al. [5]). As a result, mycorrhizal diversity may underpin many forest ecosystem services including nutrient cycling and water use efficiency. Although studies have widely demonstrated declines in ECM diversity and changes in ECM community composition following N-enrichment [3,6–12], the extent to which these changes influence the functional capacity of ECM is less well understood [13–16]. In this study, we examined the diversity and functioning in ECM communities under ambient N deposition and following N fertilization in an urban-adjacent *Pinus elliottii* plantation.

Enzymatic activities comprise one type of ECM functional trait that can be directly linked to ecosystem nutrient cycling [17]. Most ECM fungal taxa produce a diversity of extracellular and cell wall-bound hydrolytic and oxidative enzymes that mobilize the release of smaller organic molecules (potential C, N, or P sources) from soil organic matter (SOM) [17], including polyphenol–protein complexes. Studies have revealed substantial interspecific differences in ECM enzymatic activities [18–28]. Such differences can be predicted in part by ECM life history strategies. Key among these is the abundance and morphology of external hyphae among ECM taxa, also referred to as "hyphal exploration strategy" [29]. Each exploration type can vary in its capacity for enzymatic mobilization, uptake, and transfer of nutrients to the host. For example, long-distance exploration types form extensive networks of hyphae and rhizomorphs, are typically abundant in N-limited soils, and appear to be specialized in N-acquisition from complex organic substrates. Conversely, contact-types are more frequently detected in mineral soils, show lower proteolytic capabilities, and access inorganic N sources that are more readily assimilated [6,29]. Differences in ECM taxa and their exploration strategies could therefore have an impact on tree nutrition through changes in their morphological and functional (enzymatic) traits.

Biotic (host C allocation) and abiotic (climate, soil nutrients, pH) factors can also influence ECM enzyme activities. Ectomycorrhizal fungi may respond to shortages in host C allocation by up-regulating the activity of enzymes used to obtain labile carbohydrates [13], while changes in the relative availabilities of N and P are known to modify the activity of extracellular N- and P-mobilizing enzymes. For example, N fertilization could accelerate the degradation of easily decomposable litter and reduce the activity of extracellular ECM enzymes targeting recalcitrant litter with high levels of lignin and complex organic forms of N [24–28]. Both outcomes may reflect the stimulation or repression of different sets of enzymes. In addition, the activity of P-mobilizing enzymes has been shown to increase following N fertilization as a way to offset plant P demand [7]. However, neutral and negative effects have also been noted [7]. Any changes in the activity of these enzymes may reflect alterations in the ECM community and the physiological functioning of their constituent species. Such shifts, in concert with declines in ECM root colonization following N fertilization [30], could feedback to impact plant nutrient uptake.

Much of our knowledge of N enrichment effects on ECM communities has been obtained from studies in North American and European forests. However, forests in China have also experienced increasing inputs of anthropogenic N deposition owing to rapid economic development, urbanization, and intensified agricultural activities [31–34]. In the forests of south-central China, dry N deposition contributes ~24 kg N ha^{-1} year^{-1} (as NH_4-N) derived from power generation, traffic, and intensive fertilizer applications. In this region, forest plantations are comprised of a fast-growing non-native pine (*Pinus elliottii*, slash pine) that was planted to ameliorate land degradation. Although ECM fungi are critical for the growth and nutrition of *Pinus* species, it is unclear how interactions between N enrichment and a non-native pine could feedback to alter ECM communities and their ecosystem function in soil nutrient dynamics [35,36].

In this study, we examined the link between the ECM community structure and functioning under ambient N-deposition and following N fertilization in a *Pinus elliottii* (slash pine) plantation. To put our study in context with previous research, we first examined the effect of ambient N deposition and N fertilization on soil fertility and ECM community composition, diversity, and root colonization. Next, we tested the capacity of ECM fungal colonized root tips to produce an oxidative enzyme involved in the degradation of recalcitrant plant residues (phenol oxidase), and hydrolytic enzymes for organic N (protease) and P (phosphomonoesterase) mobilization. We used these results to address two questions:

(1) Are there parallel shifts in ECM fungal community structure and functioning with increasing N availability?; (2) Are there ECM fungus species-specific differences in N and P enzyme activity, and if so, do these changes reflect soil fertility or other factors (e.g., hyphal exploration)?

2. Materials and Methods

2.1. Study Site

Our study was undertaken in 35-year-old slash pine (*P. elliottii*) stands at the Hunan Forest Botanic Garden (113°02′03′ E, 28°06′07′ N), Changsha city, Hunan Province, China. The climate is typical subtropical humid monsoon with a mean annual temperature of 17.4 °C and annual precipitation of 1549 mm, most of which occurs between April and October. The soil is classed as an Alliti–Udic Ferrosol (equivalent to Acrisol; IUSS Working Group WRB, 2006), which is generally a clay loam red soil developed from slate and shale parent rock. These soils are acidic (pH = 4.14–4.21 [31]) with deficiencies of SOM and P, and high levels of Fe and Mn (Table 1).

Nine plots (each 10 m × 10 m) enclosing at least five pine trees were established in June 2010 using a completely randomized design. A 3 m buffer zone was installed around each plot to prevent N fertilizer contamination among plots. Three plots were randomly allocated to each of three N fertilizer levels: control (ambient N, no fertilization), low (50 kg N ha^{-1} year^{-1}), or high nitrogen (300 kg N ha^{-1} year^{-1}). The fertilization rates represent the expected input from N deposition in the near future (low N), as well the potential long-term cumulative N inputs from atmospheric deposition (high N [31–34]). Nitrogen fertilization treatments were applied twice a year (January, June) for three years as a solution of NH_4NO_3 uniformly sprayed across the plot. Control plots were sprayed with a similar volume of deionized water.

Table 1. Mean levels of soil nutrients in control and N-fertilized plots. Data represents mean with the standard error in parentheses.

	N Fertilization Level		
Soil nutrient	Control (*n* = 36)	Low (*n* = 36)	High (*n* = 36)
Organic matter (g kg^{-1})	26 (2) ab	21 (2) b	28 (2) a
Total N (g kg^{-1})	1.34 (0.1) a	1.26 (0.1) a	1.48 (0.1) a
Organic C:N	12 (1) a	10 (1) a	12 (2) a
Available N (µg g^{-1} soil)	26 (1.9) a	25 (1.3) a	27 (1.7) a
Total P (µg g^{-1} soil)	110 (4) b	122 (5) a	115 (4) b
Available P (µg g^{-1} soil)	3.5 (0.1) c	4.6 (0.2) a	3.0 (0.1) b
N:P	12 (1) a	11 (1) a	13 (1) a
K (µg g^{-1} soil)	69 (6.8) a	64 (2.7) a	41 (1.8) b
Ca (µg g^{-1} soil)	193 (18) a	257 (15) a	216 (16) a
Mg (µg g^{-1} soil)	896 (22) b	1051 (20) a	977 (23) a
Fe (µg g^{-1} soil)	15,986 (266) b	16,273 (140) b	17,034 (169) a
Mn (µg g^{-1} soil)	80 (7) b	101 (6) a	114 (6) a
CEC (cation exchange capacity)	20 (1.6) a	18 (1.2) a	19 (1.8) a

Means within rows with the same letter do not differ significantly at $p < 0.05$ by Tukey's Honestly Significant Difference (HSD) test.

2.2. Sample Collection

Three slash pine trees in each plot were sampled for ECM fungi in August 2013. Four soil cores (10 cm diameter, 15 cm deep), representing one core from each of the cardinal directions, were collected for each tree (total *n* = 12 cores per plot). Soil cores were placed in individual plastic bags and then stored at 4 °C until processing (within 7 days). A sub-sample of soil from each core was sieved to 2 mm and analyzed for soil N, P, K, Ca, Mg, Fe, Mn, and organic matter (OM, organic C × 1.724) at the National Engineering Laboratory for Applied Technology of Forestry and Ecology in South China. Analytical methods are detailed in supporting materials (Supplementary S1). Soil cation exchange capacity (CEC) was calculated as the sum of exchangeable cations (K, Ca, Mg) on an equivalent basis. The remaining soil in each core was sieved over 2 and 0.25 mm sieves. Fine roots collected on each sieve were gently washed to remove adhering soil and pooled for each tree. To quantify ECM root colonization, six roots (~10 cm long) per tree were randomly selected and examined by counting ECM root tip numbers. Every root tip was examined under 40× magnification for the presence of ECM colonization (i.e., turgid, swollen root tips with a well-developed mantle), and then sorted into morphological categories based on mantle color and texture, and the morphology of external hyphae on the root tip [5,29] and the publicly available database DEEMY (http://www.deemy.de/).

2.3. Enzyme Assays

Three extracellular enzymes: acid phosphatase (AP, EC 3.1.3.2), protease (PRO, EC 3.4.23), and phenol oxidase (PO, EC 1.14.18.1) were assayed. The enzyme substrates were 5 mM p-NP (*p*-nitrophenyl phosphate) for AP; 25 mM L-DOPA (L-3, 4-dihydroxyphenylalanine) for PO; and a general proteolytic substrate, Azocoll® (<50 mesh, Calbiochem-Behring Corp. La Jolla, CA, USA) for PRO [37]. All substrates were prepared in 50 mM sodium acetate-acetic acid buffer (pH 5.0, Sigma Chemical, Co. St. Louis, MO, USA). Root tip enzyme activity was assayed using the high-throughput microplate method described by Pritsch et al. [20], with one modification. Instead of a porous microplate, we constructed strips of eight plastic microscopy capsules (diameter 5 mm, height 15 mm) that were perforated at the base (Leica catalog no. 16702738); each strip of capsules fit into eight wells of a 96-well microplate.

For each classified morphotype per tree (generally 3–5 morphotypes), seven active root tips of the same diameter were selected and trimmed to 4 mm length, and one root tip was then placed in an individual capsule. Using ECM root tips with trimmed ends may have introduced intracellular enzymes into the analyses. However, hydrolytic (e.g., AP) and oxidative (e.g., PO) enzymes tend to be unaffected by cell lysis [38], and intracellular phosphatase tends to have an alkaline pH optima [39]. Non-colonized roots were not assayed for enzyme activity, as these frequently host saprophytic fungi with similar enzyme capacities as ECM [37].

Strips of capsules containing root tips were incubated in microplate wells each containing 100 μL of an individual substrate for 1 h at 37 °C (AP, PO) or two-hours (PRO) in the dark. As a control, root tips were bathed in buffer for each incubation step. At the end of each incubation period, capsules were removed from the microplate, and enzyme reactions terminated by the addition of 100 μL of sterile water (PRO, PO) or 100 μL NaOH (pH > 10, 0.2 M; AP) to each well, and absorbance measured at 405 nm (AP), 520 nm (PRO), or 450 nm (PO). After the assays were completed, each root tip was removed from the well, rinsed in deionized water, and three root tips of each morphotype per tree were frozen at −80 °C for later molecular identification, while the remaining root tips were dried to constant weight (65 °C). All measured enzyme activities were calculated per gram dry weight root per hour, and averaged among the weighted root tips of the assay group.

2.4. Identification of ECM Fungi on Root Tips

DNA from ECM root tips was extracted using DNeasy Plant Mini Kit (Qiagen SA, Coutaboeuf, France) following the manufacturer's instructions. Genomic DNA was amplified using the ITS1-F/ITS4 primer pair [40,41], after which the PCR products were visualized by gel electrophoresis. Samples with single bands were prepared for sequencing using ITS-4 and Big Dye Terminator Kit (Applied Biosystems, Foster City, CA, USA), and analyzed on an Applied Biosystems 3130xl Sequencer. PCR products with multiple bands were cloned using TOPO TA cloning kit (Invitrogen, Carlsbad, CA, USA). Successfully cloned colonies were amplified using primer pair M13F/M13R, screened using gel electrophoresis for the appropriate sized PCR products, and sequenced.

Sequences were manually aligned and edited in CodonCode Aligner 4.2.4 (CodonCode, Co. Centerville, MA, USA) and sequence homologies determined using the Basic Local Alignment Search Tool algorithm (BLAST v2.2.29 [42]) or the UNITE database v7 [43] for operational taxonomic unit

(OTU) clustering. A root tip sample was considered a species match to a database taxon if their sequences had 97% or greater similarity and were aligned over at least 450 base pairs. If no match could be made, a taxonomic placement was made by aligning the sample sequence with representative sequences of fungi from the major ECM clades. The same criteria for BLAST species matching were used to assign a taxonomic identity. Sequences from this study have been deposited in NCBI with access numbers KP866117–KP866136.

2.5. Data Analyses

All analyses—except species richness and diversity estimates—were completed in R 3.1.2 (R Project for Statistical Computing; http://www.R-project.org) with the "vegan" package [44]. Differences in the levels of soil nutrients between each treatment were analyzed using a one-way analysis of variance (ANOVA), followed by pairwise comparisons using the Tukey's Honestly Significant Difference (HSD) test. Data sets were cube root transformed before ANOVA to meet the assumptions of normality. Estimates of ECM species richness (Mao Tau, $Chao_2$, $Jackknife_2$) and diversity (Shannon-Wiener; Simpson) were calculated for each treatment in EstimateS using 50 randomizations with replacement [45]. Differences in ECM species richness and diversity and ECM root colonization among treatments were analyzed using one-way ANOVA and Tukey's HSD.

The effect of N fertilization on ECM community composition was tested using non-metric multidimensional scaling (NMDS) using Bray–Curtis dissimilarities followed by permutational analysis of variance (PERMANOVA) to test for ECM compositional differences between treatments (999 permutations). We then used vector fitting to the NMDS ordination to determine the effects of soil nutrients; significance values were generated using 999 random permutations. The effect of N fertilization on ECM community enzyme activity was similarly analyzed using NMDS and PERMANOVA.

Potential enzyme activities were used to calculate the relative contribution of each ECM species to community enzyme activity as well as changes in enzyme stoichiometry (also known as enzyme acquisition ratios) between N fertilization treatments. The relative contribution (RC) of each ECM species to community enzyme activity in each N fertilization treatment was calculated as

$$(activity_{species} \times \text{root tip abundance}_{species})/\text{total activity of the ECM community}$$

where *species* represents an individual ECM species, and total activity of the ECM community was calculated as [13,46]

$$\sum (activity \times \text{root tip abundance})_{species}.$$

Enzyme data were also used to calculate enzyme stoichiometry (also known as enzyme acquisition ratios) in each N fertilization treatment as PRO:AP, PO:PRO, and PO:AP.

Differences in potential enzyme activity (AP, PRO, PO) and stoichiometry (PRO:AP, PO:PRO, PO:AP) between N fertilization treatments and between ECM exploration and mantle types in response to N fertilization treatments were analyzed using mixed-effect ANOVA with N-treatments (control, low N, high N) and mantle type (hydrophilic, hydrophobic) or exploration type (contact, short-, and medium-distance) [5,29] as fixed effects and plots as random effects. ANOVA were followed by comparisons using Tukey's Honestly Significant Difference (HSD) test for significant variables. Relationships between the enzyme activity and ECM root colonization or soil factors were tested using Spearman's r correlation test. The relative enzyme activity in each ECM species was compared and analyzed against the community average in each enzyme and N fertilization level using *t*-tests. Data sets were square root (enzyme potential) or arcsine square root transformed (root colonization, enzyme ratios) before analyses to meet the assumptions of normality.

3. Results

3.1. Soil Fertility

Nitrogen fertilization resulted in significant increases in levels of Mn and Mg, and declines in available P and K relative to control plots. Levels of soil OM and Fe were highest in the high N fertilization treatment (Table 1). However, there was no significant effect of N fertilization on total or available soil N, C:N, total P, Ca, or CEC (18–20 ± 1.5 Meq).

3.2. ECM Community Composition and Diversity

Root tips (742; Control: 294; Low N: 224; High N: 224) were sorted into morphological groups, and from these tips, and we submitted 318 root tips for molecular analysis and successfully recovered 350 sequences, including several clones from double bands PCR products. Using a 97% sequence similarity cut-off, we identified 257 sequences representing 24 unique OTUs (hereafter referred to as species) that were ECM (Table 2). The remaining fungi were taxa traditionally considered as saprotrophic (*Paecilomyces*, *Sphaeropsis*, *Penicillium*) or of uncertain mycorrhizal status (e.g., *Basidiodendron*, *Mycena*).

Members of the Helotiales (Ascomycota) and Thelephoraceae (Basidiomycota) dominated the ECM community. Many ECM species were detected in both the control and N-fertilized plots, including *Tylospora* (Atheliaceae), *Lactarius* (Russulaceae), and members of the Ascomycota (e.g., *Helotiales* spp.). Nine ECM fungi were absent in control plots (e.g., *Scleroderma*) while an additional three taxa were N-sensitive and recovered only in the control plots (e.g., *Cenococcum* sp. 2). Levels of species richness and diversity did not differ significantly between N-fertilized and control plots (Table 3). Trees in Control (42 ± 9%) and Low N plots (44 ± 10%) showed significantly higher levels of ECM root colonization relative to high N plots (21 ± 6%; $p = 0.039$).

NMDS showed that ECM communities from N fertilization treatments were separated from one another in ordination space (Figure 1a; $p < 0.05$ for PERMANOVA among all three treatments). ECM community composition was significantly correlated with mineral nutrients (Mg, K, P) and organic C, as these resources decreased (mineral nutrients) or increased significantly (soil organic matter) in high N plots (Table 1).

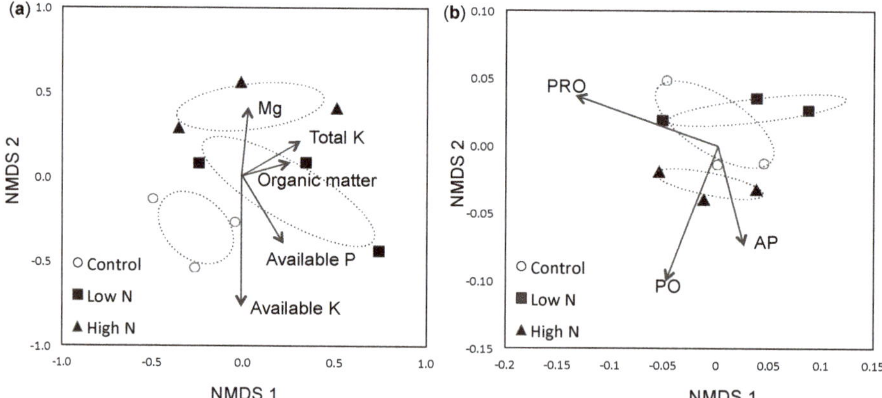

Figure 1. Non-metric multidimensional scaling (NMDS) ordination of ectomycorrhizal (ECM) fungal communities in control and N-fertilized plots based on: (**a**) the ECM root tip community, (**b**) the activity of extracellular enzymes. Each point represents the fungal community composition in each plot. Significant environmental variables are shown: P-phosphorus; K-potassium; Mg-magnesium. AP: acid phosphatase; PO: phenol oxidase; PRO: protease.

Table 2. Identification of ectomycorrhizal fungal operational taxonomic unit (OTU)s associated with *Pinus elliottii* growing at the study site in Hunan botanic garden, China. Species for group level enzyme analysis were tagged in bold font.

OTU	Accession Number [a]	Closest Blast Match in Genbank [b]	Query/Aligned length (bp) (similarity %) [c]	Closest UNITE Species Match	No. Of root tips/Frequency [d]		
					Control	Low N	High N
Tylospora **sp.**	KP866117	HM189733 Corticiaceae sp. BB-2010	632/637 (99)	SH192265.07FU	29/2	13/1	21/3
Atheliaceae **sp.**	KP866118	AB839405 Uncultured ECM fungus	475/512 (93)	SH193510.07FU	0	1/1	0
Cenococcum **sp.1**	KP866119	JQ347051 Uncultured *Cenococcum*	567/571 (99)	SH214459.07FU	7/1	28/2	0
Cenococcum **sp.2**	KP866120	JX456699 Uncultured fungus	486/497 (98)	SH214466.07FU	14/3	0	0
Helotiales **sp.1**	KP866121	KF007259 Uncultured ECM fungus	608/618 (98)	SH214286.07FU	18/4	24/5	40/6
Helotiales **sp.2**	KP866122	AB571492 Uncultured ECM fungus	637/639 (99)	SH023418.07FU	16/2	7/1	15/2
Helotiales **sp.3**	KP866123	AB769894 Uncultured Helotiales	550/551 (99)	SH201717.07FU	14/2	7/1	27/2
Helotiales sp.4	KP866124	HM208727 Fungal sp. Phylum141	560/562 (99)	SH196495.07FU	1/1	0	0
Helotiales sp.5	-	FN397286 Uncultured fungus	499/536 (93)	SH211375.07FU	0	0	1/1
Lactifluus parvigerardii	KP866125	JF975641 *Lactifluus parvigerardii* XHW-2011	571/574 (99)	SH012454.07FU	7/1	7/1	7/1
Phialocephala fortinii	KP866127	KF313098 *Phialocephala* sp. YJM2013	562/564 (99)	SH204999.07FU	1/1	0	2/1
Russula **sp.1**	KP866128	JX457011 Uncultured fungus	683/696 (98)	SH017121.07FU	0	7/1	7/1
Russula virescens	KP866129	KM373243 *Russula crustosa*	669/716 (93)	SH179774.07FU	0	1/1	0
Russula **sp.2**	KP866130	AB597671 Fungal sp. JK-02M	580/582 (99)	SH017122.07FU	0	7/1	0
Scleroderma yunnanense	KP866131	JQ639046 *Scleroderma yunnanense* XEX-2012	584/588 (100)	SH189277.07FU	1/1	0	1/1
Scleroderma citrinum	KP866132	AB769913 Uncultured *Scleroderma citrinum*	561/569 (99)	SH008294.07FU	0	0	1/1
Sebacinaceae sp.	KP866133	KF000673 Uncultured *Sebacina* clone	628/657 (96)	SH214656.07FU	0	0	1/1
Thelephora terrestris	KP866134	KJ938034 Uncultured fungus	677/706 (96)	SH184510.07FU	7/1	14/2	0
Tomentella **sp.1**	KP866135	AB769927 Uncultured Thelephoraceae	663/665 (99)	SH177859.07FU	14/2	7/1	0
Tomentella **sp.2**	KP866136	JX456648 Uncultured fungus	705/706 (99)	SH189353.07FU	37/5	2/1	7/1
Ascomycota sp.1	-	EF619719 Uncultured Orbiliaceae	525/612 (86)	SH015725.07FU	2/2	0	1/1
Ascomycota sp.2	-	KP323399 Uncultured fungus	208/226 (92)	SH469383.07FU	2/2	0	0
Ascomycota sp.3	-	KP689247 Ascomycota sp.	618/625 (99)	SH181934.07FU	0	1/1	3/3
Meliniomyces sp.	-	FJ440931 Uncultured ectomycorrhiza	553/574 (96)	SH181081.07FU	0	2/1	0

[a] Accession numbers of sequences from this study deposited in NCBI; -, sequence not deposited. [b] Closest matched BLAST results with informative species and genera. [c] Similarity values were computed from the percent match between the portion of the query aligned and its reference sequence. [d] Frequency refers to presence/absence of ECM in each focal tree and treatment (n = 9 trees per treatment).

Table 3. Estimators of operational taxonomic unit (OTU) richness and diversity of ECM fungi in different treatments. Data represents mean with the standard error in parentheses per plot; $n = 3$ (3).

Sites	Rarified Species Richness		Estimators of Expected Total Species Richness		Diversity Indices	
	Mao Tau	**Mao Tau (50 runs mean)**	**Chao 2**	**Jackknife 2**	**Shannon's *H'***	**Simpson's 1/D**
Control	5.67 (1.67)	6.41 (1.58)	8.54 (1.20)	8.78 (2.19)	2.79 (0.96)	5.89 (1.31)
Low N	4.33 (1.45)	4.5 (1.08)	5.1 (1.24)	6.28 (1.53)	1.35 (0.25)	4.06 (0.98)
High N	4.33 (0.67)	5.1 (0.37)	5.99 (0.32)	6.80 (0.12)	2.84 (1.40)	4.59 (0.43)
p [a]	0.729	0.508	0.119	0.52	0.519	0.444

[a] p-value for effect of fertilizer treatment.

3.3. Extracellular Enzyme Activity

Although we measured enzyme activities in 742 individual ECM root tips, we used results from only those ECM taxa with well-supported molecular identities in the statistical and comparative analyses. Consequently, we used the results from 426 root tips (Control: 166; Low N: 126; High N: 134), which represented eight ECM fungal taxa. These taxa were recovered in sufficient numbers across all treatments so that at least three individual root tips of each species from each treatment could be assayed (Table 2, species names in bold). Overall, High N fertilization resulted in a significant increase in PO activity ($p = 0.004$ for ANOVA) and decrease in AP activity ($p = 0.035$ for ANOVA), but had no significant effect on PRO activity ($p = 0.795$ for ANOVA) (Figure 2).

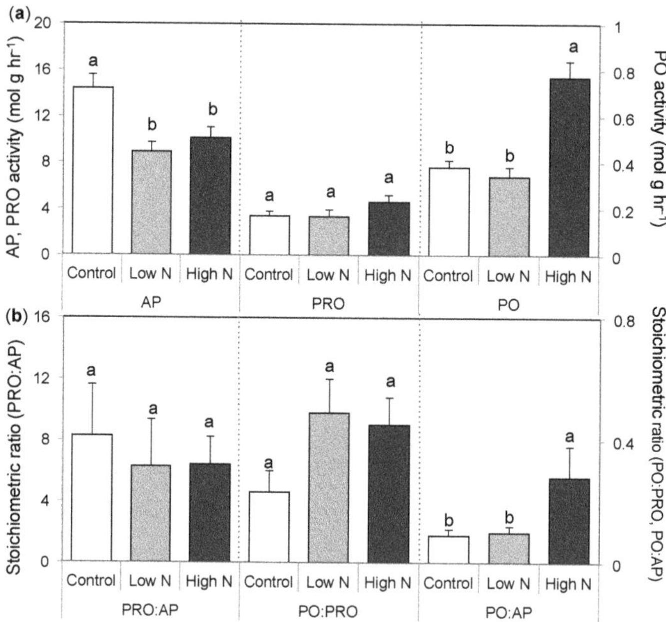

Figure 2. Mean levels of (**a**) acid phosphatase, protease, and phenol oxidase activity; and (**b**) enzymatic stoichiometric PRO:AP, PO:PRO, and PO:AP in ectomycorrhizal communities in control and N-fertilized plots. Vertical bars indicate the standard error of the mean; for each enzyme (or ratio), columns with the same letter do not differ significantly at $p < 0.05$ based on Tukey's HSD test.

Potential AP and PO activity were correlated with levels of soil P and K. Potential AP activity was negatively correlated with total ($r = -0.086$, $p = 0.023$) and available soil P ($r = -0.077$, $p = 0.044$), and total K ($r = -0.082$, $p = 0.030$). Potential PO activity was positively correlated with soil P ($r = 0.154$, $p < 0.001$) and available ($r = 0.109$, $p = 0.004$) and total K ($r = 0.101$, $p = 0.008$). There was no relationship between PRO and any tested soil factor. Potential enzyme activity was also correlated with root tip colonization in low ($r = 0.611$, $p = 0.001$) and high N plots ($r = 0.445$, $p = 0.029$), but not in control plots ($r = 0.229$, $p = 0.281$).

The NMDS ordination showed significant differences in ECM community between N fertilization treatments based on enzyme activity (Figure 1b; $p = 0.001$ for PERMANOVA). This pattern was driven by PO activity, because high N plots had significantly higher levels of PO activity than control or low N plots (Figure 2).

Enzyme activity varied significantly between ECM species (Figure 3; Figures S1 and S2). Overall, *Helotiales* sp.1, and Thelephoraceae were the largest contributors to ECM community enzyme activity in most enzyme systems and N fertilizer treatments, and relative enzyme activity in these taxa was always significantly greater than the community mean (Figure 3a–c). Certain taxa were restricted to a specific enzyme system or N fertilization treatment (e.g., *Lactarius* for PO in N plots; Figure 3c), but for the most part, ECM taxa varied in their relative activity among enzyme systems and N fertilization treatments. For example, the activity of Atheliaceae has greater contribution to the community for AP in high N plots (Figure 3a), and for PRO in control plots (Figure 3b). There was also strong inter-specific variation in relative enzyme activity among the three species of Helotiales.

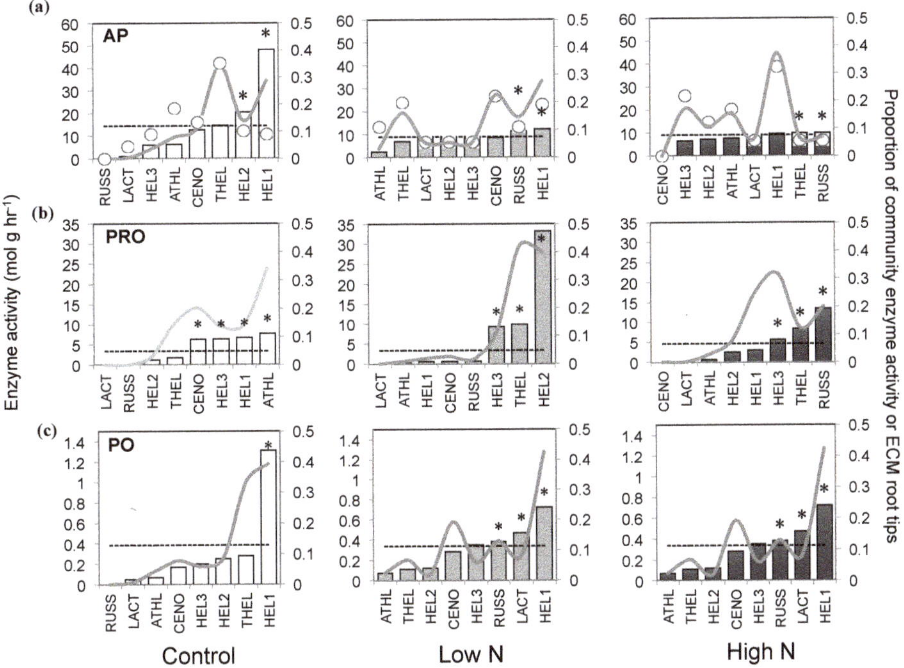

Figure 3. Levels of ectomycorrhizal (ECM) root tip relative abundance (open dot) and the relative contribution (grey curve) of individual ECM group to overall ECM community activity of (**a**) acid phosphatase (AP), (**b**) protease (PRO), and (**c**) phenol oxidase (PO) in response to N fertilization. Within each panel for enzyme and N fertilization treatment, broken lines within each panel indicate the mean community level of enzyme activity. Columns denoted with an asterisk (*) denote ECM species in which enzyme activity was significantly greater ($p < 0.05$) than the mean community value.

Fertilization altered the stoichiometry of enzyme activity by significantly increasing the depletion of AP (P-cycling) relative to PO and PRO (C-, N-cycling, respectively; Figure 2b). In N-fertilized plots, contact (Russulaceae) and medium-fringe (Atheliaceae) types showed a significant increase in the PRO:AP ratio, whereas short (*Cenococcum* and Helotiales) and medium-smooth (Thelephoraceae) types showed a decline (Figure 4a). Conversely, the PO:AP ratio increased in medium-smooth types and declined in medium-fringe types (Figure 4c). The PO:PRO ratio declined significantly (contact, medium-distance types) or did not differ significantly between treatments (short-distance; Figure 4b). Although the PO:AP ratio was higher in ECM species with hydrophilic mantles (groups of *Cenococcum*, Thelephoraceae, Russulaceae, and *Helotiales* spp.) compared to those with hydrophobic mantles (group of *Atheliaceae* spp.) (Figure 5), this difference was not statistically significant ($p = 0.223$). Similarly, PO:AP activity was greatest in high N plots, but again, this difference was not statistically significant (hydrophilic $p = 0.746$; hydrophobic $p = 0.391$).

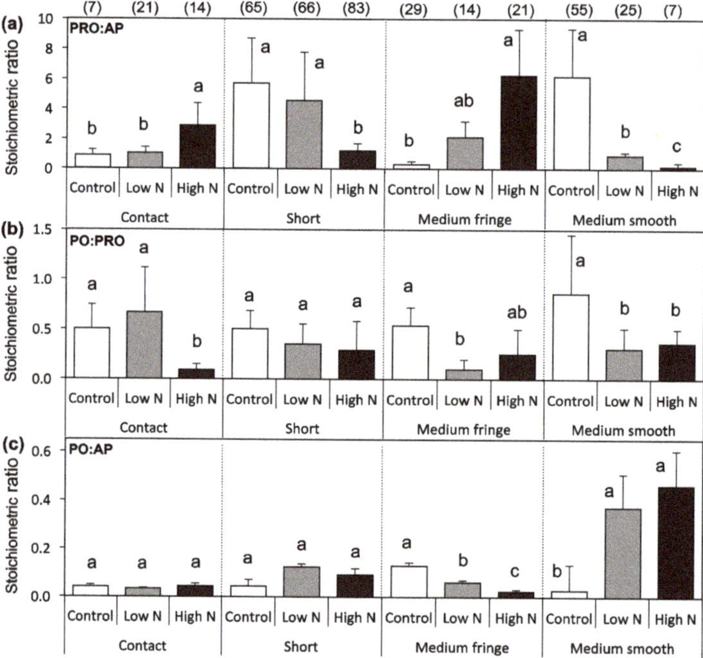

Figure 4. Mean (**a**) PRO:AP, (**b**) PO:PRO, and (**c**) PO:AP in ectomycorrhizal communities in control and N-fertilized plots based on hyphal exploration strategy. For each enzyme and hyphal strategy, columns with the same letter do not differ significantly at $p < 0.05$ based on Tukey's HSD test. Root tip number per exploration type is listed on the top panel.

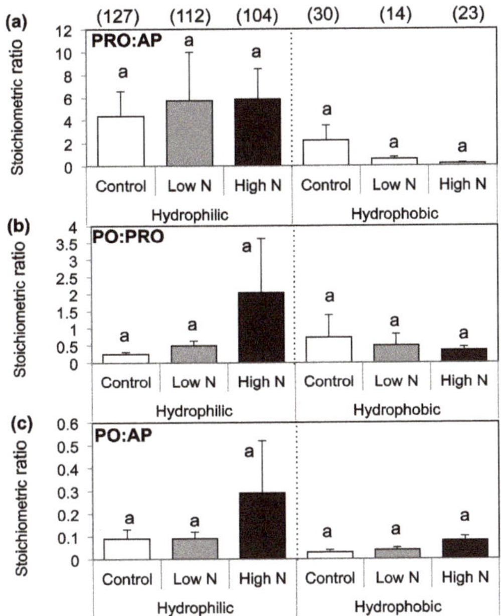

Figure 5. Mean (**a**) PRO:AP, (**b**) PO:PRO, and (**c**) PO:AP ratios based on mantle structure in control and N-fertilized plots. Mantle types with the same letter did not differ significantly at $p < 0.05$ based on Tukey's HSD test. Root tip number per hydrophobicity type is listed on the top panel.

4. Discussion

Despite the high levels of ambient N deposition, we found that three years of N fertilization produced shifts in ECM fungal community structure, and variously altered the abundances and potential enzyme activities of ECM fungal taxa as well as the stoichiometry of extracellular enzymes involved in P acquisition (AP) and lignin degradation (PO) (Question 1). Such shifts were related to some extent to changes in nutrient availability (P, K, OM) created by N fertilization (Question 2) and differences in enzymatic stoichiometry among ECM hyphal exploration types.

ECM fungal richness, diversity, and evenness were not significantly different across N fertilization treatments. These results reconcile with earlier studies in oak [30] and spruce forests [3,47–49], but seemingly contradict the majority of studies that show a rapid and substantial decline in ECM fungal diversity following N fertilization e.g., [6,9–11,30] and increasing soil N levels along natural gradients of productivity [50]. This result may reflect the short period (three years) over which the plots were fertilized [6,51] or (more likely) that high levels of ambient N deposition may have pre-empted any effects of the experimental N additions. Indicative of this condition, we found that even high N fertilization (300 kg N ha^{-1} year^{1}) did not significantly increase the levels of total and available N over those in non-fertilized plots, and the levels of available N across all plots (25–27 mg N kg^{-1} soil) exceeded those documented in an N-saturated pine forest in southern China [31]. Further, N additions are expected to alter the quantity and quality of resources (i.e., litter C:N) so as to influence soil C:N. However, we found that the soil C:N ratio did not differ significantly between N-fertilized and control (ambient N) plots. Taken together, these findings suggest the soil was N-saturated so that any new input of N (i.e., fertilizer) was likely leached from the system.

Even so, N fertilization was sufficient to alter ECM fungal community composition. The major genera found in our study (e.g., *Cenococcum*, *Thelephora*, *Tomentella*, *Sebacina*, *Inocybe*, *Russula*) are ubiquitous and dominant components in ECM communities. The most obvious indicators of

N-enrichment were the increased abundance of ECM taxa considered to be nitrophilic (*Russula, Tomentella*), the loss of N-sensitive fungi (*Cenococcum*), and the exclusion of protein-N-utilizing ECM species (*Suillus, Tricholoma, Cortinarius, Piloderma* [5,30,48,49]). These results also reflect the shift to an ECM fungal community comprising contact- (Russulaceae) and short-range exploration types (*Cenococcum*, Helotiales), and certain medium-distance types (Thelephoraceae, Atheliaceae). Our findings are in agreement with the well-documented patterns of ECM community change and species' abundances in response to N-deposition or fertilization noted elsewhere [8,11,30,49]. Unlike these former studies, soil K, P, and OM were the primary soil factors associated with changes in ECM fungal community structure, not N. This distinction could reflect the difference in major N input as NH_4-N versus NO_3-N in our systems. Nevertheless, our results show that imbalances of mineral nutrients will become increasingly important controls over ECM fungal community structure in ecosystems with chronic N deposition or saturation.

Forest soils can become increasingly deficient in K and P as a result of atmospheric N deposition or N fertilization, as occurred in our study plots [1]. However, ECM fungal communities respond differentially to deficiencies of K versus P. For instance, deficiencies of soil K reduce fine root growth and ECM colonization by impeding sucrose export from the leaves to the roots [52,53], but have no effect on ECM community composition [54]. Conversely, P-deficiency tends to increase C allocation belowground to stimulate root and ECM fungal growth and AP activity [55]. In our study, the decrease in ECM root colonization with N fertilization is consistent with K deficiency. In contrast, the decline in AP activity and changes in enzyme stoichiometry toward an increase in C-acquisition (PO) relative to P (AP) with N fertilization was a clear contradiction of previous studies [56]. The precise reason(s) for this response are beyond the scope of this study. However, it is possible that AP activity on ECM root tips may have been low relative to the activity in the foraging extraradical mycelium in other parts of the soil profile (i.e., functional compartmentalization [22,55]) or that there were changes in the physiological competence of ECM root tips owing to reductions in C-allocation from the host [52,53].

N fertilization also resulted in ECM communities with high PO activity. N-fertilized communities showed a net up-regulation in PO activity as individual ECM fungi showed an increased capacity to degrade SOM. This is in general agreement with the widespread capacity of ECM fungi to oxidize SOM [21,57,58] and the generally positive effect of soil N on litter degradation by ECM fungi [2,7]. Such increases in PO activity have been interpreted as a mechanism by which ECM acquire labile carbohydrates when host C allocation is low [22,58], or that ECM generally degrade polyphenolic-rich compounds as a consequence of mining for nutrients [20,22,59]. However, these mechanisms are not mutually exclusive, and may vary in their importance depending on the extent of N-enrichment and/or mineral nutrient deficits. Further studies are thus needed to determine the precise role(s) of each mechanism in this system. Even so, the positive relationship between PO and AP activity suggests that ECM fungi may have been similarly prospecting for P. Similarly, the significant correlations between PO and soil P, K, and Mg content are consistent with the concept that ECM use oxidative enzymes such as PO to mobilize mineral nutrients locked within organic matter-mineral complexes [13,17,20,24,57].

Without examining the relationship between ECM enzyme levels and chemical modifications of SOM, we cannot exclude the possibility that ECM fungi also acquired labile C from degraded substrates. In our analyses of ECM enzyme stoichiometry and potential energetic trade-offs, only ECM fungi with the medium-distance smooth exploration type (Thelephoraceae) showed an increase in SOM-degrading potential relative to P acquisition. This is not surprising, since the Thelephoraceae tend to dominate in organic-rich soil horizons [18,60] and show some of the strongest potential activities of degradative enzymes [4,6,61,62]. In addition, studies have shown that when host plants allocate less C to their ECM fungi, either due to dormancy or defoliation, the relative activity of C-degrading enzymes increases significantly [17,55,63].

ECM fungi may also experience trade-offs between the energetic demands associated with root tip colonization versus the metabolically expensive production of enzymes. Under ambient N conditions (control), ECM species with lower levels of root colonization showed disproportionately high levels of potential enzyme activity—an outcome that is consistent with energetic trade-offs [46]. Communities in N-fertilized plots displayed a different pattern, whereby root tip abundance was positively associated with enzyme activity (Figure 3). Such results are more typical of competitive interactions; that is, where certain ECM taxa pre-emptively colonized roots and utilized resources [64,65] owing to competition for spatial co-existence (a limited availability of root tips) or overlap in resource utilization (limited C availability). Alternatively, the abiotic conditions created by N fertilization may have selected for more closely-related taxa than expected by chance (environmental filtering [62]; e.g., members of the Russulaceae and Thelephoraceae dominated N-fertilized plots). However, additional studies are needed to distinguish between these explanations.

There is now substantial evidence that labile-C compounds derived from roots and fungal tissues are the dominant inputs into stable SOC stores [66]. Up to 30% of the total C assimilated by plants may be transferred to their ECM partner, meaning that any change in the symbiosis will have profound effects on SOC storage. Our results address two sets of dynamics relating SOC to N availability. The first focuses on the dynamics associated with increasing potential enzyme activity for SOM degradation and the release of labile C compounds. Such elevated inputs of labile C may concomitantly stimulate microbial activity and SOC decomposition (the "priming effect" hypothesis). These changes would be expected to be associated with a more rapid turnover of the SOC pool and changes in litter chemistry. However, more rapid decomposition is not synonymous with reductions in total SOC stocks if coupled with similar increases in litter inputs and soil C stabilization. Conversely, soil microbes may preferentially metabolize these pools of labile C over the mining of complex polymers (the "preferential substrate utilization" hypothesis). This effect, combined with the observations that N fertilization can inhibit SOC decomposition [1,2], may lead to increases in SOC following N fertilization.

The second focuses on ECM community dynamics. Carbon transferred to ECM fungi can also contribute to SOC storage if fungal tissues decompose more slowly than non-mycorrhizal roots [67] or if the fungal residues persist in long-term SOC stores. As a corollary, the tissue quality (C:N, melanin content) of fungal necromass can influence degradation dynamics [68]. Thus, shifts in ECM community composition and root colonization—and especially the differences in the extent of root tip colonization between different ECM fungal species and their tissue chemistry—could feedback to influence the quantity and quality of fungal residues entering the long-term C pool [69].

In the near future, this region in China is predicted to experience increasing N deposition. Based on our results, we can hypothesize that increasing inputs of N are likely to exacerbate soil P and K deficiencies, and compromise the capacity of ECM fungal communities to acquire mineral nutrients. In addition, variations in ECM community composition and species' functional plasticity could undermine the contributions of fungal residues to long-term SOC stores. These findings add to the accumulating evidence that increasing inputs of N—either from atmospheric deposition or fertilization—will continue to impact forest health and productivity by altering soil mineral resources and ECM community structure and functioning. Such prospects point to the need for a better understanding of the role(s) of ECM functional traits, their interactions with host plant growth and nutrient status, and their link to the relative soil nutrient availabilities. This multi-faceted approach is urgently needed to improve forest health and productivity in this region, as well as patterns of C stabilization and loss.

Supplementary Materials: The following are available online at www.mdpi.com/1999-4907/9/3/99/s1; Supplementary S1: details of nutrient analyses. Figure S1: potential AP, PO, and PRO enzyme activity in ECM hyphal exploration types in response to N fertilization. Figure S2: potential AP, PO, and PRO enzyme activity in ECM hyphal hydrophobicity types in response to N fertilization.

Acknowledgments: The Special Scientific Research Fund of Forestry Public Welfare Profession of China (Grant No.200804030) and the Chicago Botanic Garden provided financial support for this study. We gratefully acknowledge the in-kind support of National Engineering Laboratory for Applied Technology of Forestry and Ecology in South China, Central South University of Forestry and Technology, Changsha, and additional field and lab assistance provided by Huizhao Luo and Luyan. Xu. We appreciated valuable comments and insights from Dr. Peter Avis, and the anonymous reviewers.

Author Contributions: Chen Ning, Gregory M. Mueller, Louise M. Egerton-Warburton and Andrew W. Wilson conceived and designed the experiments; Chen Ning performed the experiments; Chen Ning and Louise M. Egerton-Warburton analyzed the data; Gregory M. Mueller, Louise M. Egerton-Warburton, Wende Yan and Wenhua Xiang contributed reagents/materials/analysis tools; Chen Ning and Louise M. Egerton-Warburton drafted the manuscript and all authors contributed to manuscript revision.

Conflicts of Interest: The authors declare no conflict of interest.

References

1. Galloway, J.N.; Dentener, F.J.; Capone, D.G.; Boyer, R.W.; Howarth, S.P.; Seitzinger, G.P.; Asner, C.C.; Cleveland, P.A.; Green, E.A.; Holland, D.M.; et al. Nitrogen cycles: Past, present, and future. *Biogeochemistry* **2004**, *70*, 153–226. [CrossRef]
2. Carreiro, M.M.; Sinsabaugh, R.L.; Repert, D.A.; Parkhurst, D.F. Microbial enzyme shifts explain litter decay responses to simulated nitrogen deposition. *Ecology* **2000**, *81*, 2359–2365. [CrossRef]
3. Fransson, P.M.; Taylor, A.F.S.; Finlay, R.D. Effects of continuous optimal fertilization on belowground ectomycorrhizal community structure in a Norway spruce forest. *Tree Physiol.* **2000**, *20*, 599–606. [CrossRef] [PubMed]
4. Dupra, C.; Stevens, C.J.; Ranke, T.; Bleeker, A.; Peppler-Lisbach, C.; Gowing, D.J.G.; Dise, N.B.; Dorland, E.; Bobbink, R.; Diekmann, M. Changes in species richness and composition in European acidic grasslands over the past 70 years: the contribution of cumulative atmospheric nitrogen deposition. *Glob. Chang. Biol.* **2000**, *16*, 3443–3457. [CrossRef]
5. Lilleskov, E.A.; Hobbie, E.A.; Horton, T.R. Conservation of ectomycorrhizal fungi: Exploring the linkage between functional and taxonomic responses to anthropogenic N deposition. *Fungal Ecol.* **2011**, *4*, 174–183. [CrossRef]
6. Lilleskov, E.A.; Fahey, T.J.; Horton, T.R.; Lovett, G. Belowground ectomycorrhizal fungal community change over a nitrogen deposition gradient in Alaska. *Ecology* **2002**, *83*, 104–115. [CrossRef]
7. Jones, M.D.; Phillips, L.A.; Treu, R.; Ward, V.; Berch, S.M. Functional responses of ectomycorrhizal fungal communities to long-term fertilization of lodgepole pine (*Pinus contorta* Dougl. ex Loud. var. *latifolia* Engelm.) stands in central British Columbia. *Appl. Soil. Ecol.* **2012**, *60*, 29–49.
8. Avis, P.G.; McLaughlin, D.J.; Dentinger, B.C.; Reich, P.B. Long-term increase in nitrogen supply alters above- and below- ground ectomycorrhizal communities and increases the dominance of *Russula* spp. in a temperate oak savanna. *New Phytol.* **2003**, *160*, 239–253. [CrossRef]
9. Högberg, M.; Blaško, R.; Bach, L.H.; Hasselquist, N.J.; Egnell, G.; Näsholm, T.; Högberg, P. The return of an experimentally N-saturated boreal forest to an N-limited state: observations on the soil microbial community structure, biotic N retention capacity and gross N mineralization. *Plan. Soil.* **2014**, *381*, 45–60. [CrossRef]
10. Parrent, J.L.; Vilgalys, R. Biomass and compositional responses of ectomycorrhizal fungal hyphae to elevated CO_2 and nitrogen fertilization. *New Phytol.* **2007**, *176*, 164–174. [CrossRef] [PubMed]
11. Cox, F.; Barsoum, N.; Lilleskov, E.A.; Bidartondo, M.I. Nitrogen availability is a primary determinant of conifer mycorrhizas across complex environmental gradients. *Ecol. Lett.* **2010**, *13*, 1103–1113. [CrossRef] [PubMed]
12. Morrison, E.W.; Frey, S.D.; Sadowsky, J.J.; Van Diepen, L.T.A.; Thomas, W.K.; Pringle, A. Chronic nitrogen additions fundamentally restructure the soil fungal community in a temperate forest. *Fungal Ecol.* **2016**, *23*, 48–57. [CrossRef]
13. Courty, P.E.; Franc, A.; Garbaye, J. Temporal and functional pattern of secreted enzyme activities in an ectomycorrhizal community. *Soil Biol. Biochem.* **2010**, *42*, 2022–2025. [CrossRef]
14. Courty, P.E.; Labbe, J.; Kohler, A.; Marcais, B.; Bastien, C.; Churin, J.L.; Garbaye, J.; Le Tacon, F. Effect of poplar genotypes on mycorrhizal infection and secreted enzyme activities in mycorrhizal and non-mycorrhizal roots. *J. Exp. Bot.* **2011**, *62*, 249–260. [CrossRef] [PubMed]
15. Jones, M.D.; Twieg, B.D.; Ward, V.; Barker, J.; Durall, D.M.; Simard, S.W. Functional complementarity of Douglas-fir ectomycorrhizas for extracellular enzyme activity after wildfire or clearcut logging. *Funct. Ecol.* **2010**, *24*, 1139–1151. [CrossRef]
16. Herzo, C.; Peter, M.; Pritsch, K.; Günthardt-Goerg, M.S.; Egli, S. Drought and air warming affects abundance and exoenzyme profiles of *Cenococcum geophilum* associated with *Quercus robur*, *Q. petraea* and *Q. pubescens*. *Plant Biol.* **2013**, *15*, 230–237. [CrossRef] [PubMed]

17. Courty, P.E.; Bréda, N.; Garbaye, J. Relation between oak tree phenology and the secretion of organic matter degrading enzymes by *Lactarius quietus* ectomycorrhizas before and during bud break. *Soil Biol. Biochem.* **2007**, *39*, 1655–1663. [CrossRef]

18. Buée, M.; Courty, P.E.; Mignot, D.; Garbaye, J. Soil niche effect on species diversity and catabolic activities in an ectomycorrhizal fungal community. *Soil Biol. Biochem.* **2007**, *39*, 1947–1955. [CrossRef]

19. Courty, P.E.; Pritsch, K.; Schloter, M.; Hartmann, A.; Garbaye, J. Activity profiling of ectomycorrhiza communities in two forest soils using multiple enzymatic tests. *New Phytol.* **2005**, *167*, 309–319. [CrossRef] [PubMed]

20. Pritsch, K.; Garbaye, J. Enzyme secretion by ECM fungi and exploitation of mineral nutrients from soil organic matter. *Ann. For. Sci.* **2011**, *68*, 25–32. [CrossRef]

21. Luis, P.; Kellner, H.; Zimdars, B.; Langer, U.; Martin, F.; Buscot, F. Patchiness and spatial distribution of laccase genes of ectomycorrhizal, saprotrophic, and unknown basidiomycetes in the upper horizons of a mixed forest cambisol. *Microb. Ecol.* **2005**, *50*, 570–579. [CrossRef] [PubMed]

22. Talbot, J.M.; Bruns, T.D.; Smith, D.P.; Branco, S.; Glassman, S.I.; Erlandson, S.; Vilgalys, R.; Peay, K.G. Independent roles of ectomycorrhizal and saprotrophic communities in soil organic matter decomposition. *Soil Biol. Biochem.* **2013**, *57*, 282–291. [CrossRef]

23. Kohler, A.; Kuo, A.; Nagy, L.G.; Morin, E.; Barry, E.W.; Buscot, F.; Canbäck, B.; Choi, C.; Cichocki, N.; Clum, A.; et al. Convergent losses of decay mechanisms and rapid turnover of symbiosis genes in mycorrhizal mutualists. *Nat. Genet.* **2015**, *47*, 410–415. [CrossRef] [PubMed]

24. Saiya-Cork, K.R.; Sinsabaugh, R.L.; Zak, D.R. The effects of long term nitrogen deposition on extracellular enzyme activity in an *Acer saccharum* forest soil. *Soil Biol. Biochem.* **2002**, *34*, 1309–1315. [CrossRef]

25. Frey, S.D.; Knorr, M.; Parrent, J.L.; Simpson, R.T. Chronic nitrogen enrichment affects the structure and function of the soil microbial community in temperate hardwood and pine forests. *For. Ecol. Manag.* **2004**, *196*, 159–171. [CrossRef]

26. Blackwood, C.B.; Waldrop, M.P.; Zak, D.R.; Sinsabaugh, R.L. Molecular analysis of fungal communities and laccase genes in decomposing litter reveals differences among forest types but no impact of nitrogen deposition. *Environ. Microbiol.* **2007**, *5*, 1306–1316. [CrossRef] [PubMed]

27. Lucas, R.W.; Casper, B.B. Ectomycorrhizal community and extracellular enzyme activity following simulated atmospheric N deposition. *Soil Biol. Biochem.* **2008**, *40*, 1662–1669. [CrossRef]

28. Eaton, G.K.; Ayres, M.P. Plasticity and constraint in growth and protein mineralization of ectomycorrhizal fungi under simulated nitrogen deposition. *Mycologia* **2002**, *94*, 921–932. [CrossRef] [PubMed]

29. Agerer, R. Fungal relationships and structural identity of their ectomycorrhizae. *Mycol. Prog.* **2006**, *5*, 67–107. [CrossRef]

30. Avis, P.G.; Mueller, G.M.; Lussenhop, J. Ectomycorrhizal fungal communities in two North American oak forests respond to nitrogen addition. *New Phytol.* **2008**, *179*, 472–483. [CrossRef] [PubMed]

31. Wen, L.; Lei, P.; Xiang, W.; Yan, W.; Liu, S. Soil microbial biomass carbon and nitrogen in pure and mixed stands of Pinus massoniana and Cinnamomum camphora differing in stand age. *For. Ecol. Manag.* **2014**, *328*, 150–158. [CrossRef]

32. Du, C.Y.; Zeng, G.M.; Zhang, G.; Tang, L.; Li, X.D.; Huang, D.L.; Huang, L.; Jiang, Y.M. Input-output budgets for inorganic nitrogen under acid rain in a subtropical evergreen mixed forest in central-south China. *Water Air Soil Pollut.* **2008**, *190*, 171–181. [CrossRef]

33. Fang, Y.; Gundersen, P.; Vogt, R.D.; Koba, K.; Chen, F.; Chen, X.Y.; Yoh, M. Atmospheric deposition and leaching of nitrogen in Chinese forest ecosystems. *J. For. Res.* **2011**, *16*, 341–350. [CrossRef]

34. Liu, L.; Zhang, X.; Wang, S.; Lu, X.; Ouyang, X. A review of spatial variation of inorganic nitrogen (N) wet deposition in China. *PLoS ONE* **2016**, *11*, e0146051. [CrossRef] [PubMed]

35. Chapela, I.H.; Osher, L.J.; Horton, T.R.; Henn, M.R. Ectomycorrhizal fungi introduced with exotic pine plantations induce soil carbon depletion. *Soil. Biol. Biochem.* **2001**, *32*, 1733–1740. [CrossRef]

36. Bastias, B.A.; Anderson, I.C.; Xu, Z.; Cairney, J.W.G. RNA- and DNA-based profiling of soil fungal communities in a native Australian eucalypt forest and adjacent *Pinus elliotti* plantation. *Soil Biol. Biochem.* **2007**, *39*, 3108–3114. [CrossRef]

37. Colpaert, J.; Van Laere, A. A comparison of the extracellular enzyme activities of two ectomycorrhizal and a leaf-saprotrophic basidiomycete colonizing beech leaf litter. *New Phytol.* **1996**, *134*, 133–141. [CrossRef]

38. Blankinship, J.C.; Becerra, C.A.; Schaeffer, S.M.; Schinmer, J.P. Separating cellular metabolism from exoenzyme activity in soil organic matter decomposition. *Soil. Biol. Biochem.* **2014**, *71*, 68–75. [CrossRef]

39. Plassard, C.; Louche, J.; Ali, M.A.; Duchemn, M.; Legname, E.; Cloutier-Hurteau, B. Diversity in phosphorus mobilization and uptake in ectomycorrhizal fungi. *Ann. For. Sci.* **2011**, *68*, 33–43. [CrossRef]

40. White, T.J.; Bruns, T.; Lee, S.; Taylor, J. Amplification and Direct Sequencing of Fungal Ribosomal RNA Genes for Phylogenetics. In *PCR Protocols: A Guide to Methods and Applications*; Innis, M.A., Gelfand, D.H., Sninsky, J.J., White, T.J., Eds.; Academic Press, Inc.: San Diego, CA, USA, 1990; pp. 315–322.

41. Gardes, M.; Bruns, T.D. Community structure of ectomycorrhizal fungi in a *Pinus muricata* forest: Above- and below-ground views. *Can. J. Bot.* **1996**, *74*, 1572–1583. [CrossRef]

42. Altschul, S.F.; Gish, W.; Miller, W.; Myer, E.W.; Lipman, D.J. Basic local alignment search tool. *J. Mol. Biol.* **1990**, *215*, 403–410. [CrossRef]

43. Kõljalg, U.; Nilsson, R.H.; Abarenkov, K.; Tedersoo, L.; Taylor, A.F.S.; Bahram, M.; Bates, S.T.; Bruns, T.D.; Pengtsson-Palme, J.; Callaghan, T.M.; et al. Towards a unified paradigm for sequence-based identification of fungi. *Mol. Ecol.* **2013**, *22*, 5271–5277. [CrossRef] [PubMed]

44. Oksanen, J.; Blanchet, F.G.; Friendly, M.; Kindt, R.; Legendre, P.; McGlinn, D.; Minchin, P.R.; O'Hara, R.B.; Simpson, G.L.; Solymos, P.; et al. Vegan: Community Ecology Package Version 2.4-4. Available online: https://github.com/vegandevs/vegan (accessed on 24 August 2017).

45. Colwell, R.K. EstimateS: Statistical Estimation of Species Richness and Shared Species from Samples. Version 9. Available online: http://purl.oclc.org/estimates (assessed on 12 August 2016).

46. Walker, J.K.M.; Cohen, H.; Higgins, L.M.; Kennedy, P.G. Testing the link between community structure and function for ectomycorrhizal fungi involved in a global tripartite symbiosis. *New Phytol.* **2014**, *202*, 287–296. [CrossRef] [PubMed]

47. Hay, T.N.; Phillips, L.A.; Nicholson, B.A.; Jones, M.D. Ectomycorrhizal community structure and function in interior spruce forests of British Columbia under long-term fertilization. *For. Ecol. Manag.* **2015**, *350*, 87–95. [CrossRef]

48. Berch, S.M.; Brockley, R.P.; Battigelli, J.P.; Haerman, S.; Holl, B. Impacts of repeated fertilization on components of the soil biota under a young lodgepole pine stand in the interior of British Columbia. *Can. J. For. Res.* **2006**, *36*, 1415–1426. [CrossRef]

49. Peter, M.; Ayer, F.; Egli, S. Nitrogen addition in a Norway spruce stand altered macromycete sporocarp production and below-ground ectomycorrhizal species composition. *New Phytol.* **2001**, *149*, 311–325. [CrossRef]

50. Kranabetter, J.M.; Durall, D.; MacKenzie, W. Diversity and species distribution of ectomycorrhizal fungi along productivity gradients of a southern boreal forest. *Mycorrhiza* **2009**, *19*, 99–111. [CrossRef] [PubMed]

51. Hasselquist, N.; Högberg, P. Dosage and duration effects of nitrogen additions on ectomycorrhizal sporocarp production and functioning: An example from two N-limited boreal forests. *Ecol. Evol.* **2014**, *4*, 3015–3026. [CrossRef] [PubMed]

52. Garcia, K.; Zimmerman, S.D. The role of mycorrhizal associations in plant potassium nutrition. *Front. Plant Sci.* **2014**, *5*, 337. [CrossRef] [PubMed]

53. Gagnon, B.; Bélanger, G.; Nolin, M.C.; Simard, R. Relationships between soil cations and plant characteristics based on spatial variability in a forage field. *Can. J. Plant Sci.* **2003**, *83*, 343–350. [CrossRef]

54. Wang, L.; Katzensteiner, K.; Schume, H.; van Loo, M.; Godbold, D.L. Potassium fertilization affects the distribution of fine roots but does not change ectomycorrhizal community structure. *Ann. For. Sci.* **2016**, *73*, 691–702. [CrossRef]

55. Finlay, R. Ecological aspects of mycorrhizal symbiosis: With special emphasis on the functional diversity of interactions involving the extraradical mycelium. *J. Exp. Bot.* **2008**, *59*, 1115–1126. [CrossRef] [PubMed]

56. Olander, L.P.; Vitousek, P.M. Regulation of soil phosphatase and chitinase activity by N and P availability. *Biogeochem.* **2000**, *49*, 175–190. [CrossRef]

57. Phillips, L.A.; Ward, V.; Jones, M.D. Ectomycorrhizal fungi contribute to soil organic matter cycling in sub-boreal forests. *ISME J.* **2014**, *8*, 699–713. [CrossRef] [PubMed]

58. Courty, P.E.; Franc, A.; Pierrat, J.C.; Garbaye, J. Temporal changes in the ectomycorrhizal community in two soil horizons of a temperate oak forest. *Appl. Environ. Microbiol.* **2008**, *74*, 5792–5801. [CrossRef] [PubMed]

59. Rineau, F.; Courty, P. Secreted enzymatic activities of ectomycorrhizal fungi as a case study of functional diversity and functional redundancy. *Ann. For. Sci.* **2011**, *68*, 69–80. [CrossRef]

60. Tedersoo, L.; Kõljalg, U.; Hallenberg, N.; Larsson, K.-H. Fine scale distribution of ectomycorrhizal fungi and roots across substrate layers including coarse woody debris in a mixed forest. *New Phytol.* **2003**, *159*, 153–165. [CrossRef]
61. Tedersoo, L.; Bahram, M.; Toots, M.; Diédhiou, A.G.; Henkel, T.W.; Kjøller, R.; Morris, M.H.; Nara, K.; Nouhra, E.; Peay, K.G.; et al. Towards global patterns in the diversity and community structure of ectomycorrhizal fungi. *Mol. Ecol.* **2012**, *21*, 4160–4170. [CrossRef] [PubMed]
62. Courty, P.; Munoz, F.; Selosse, M.-A.; Duchemin, M.; Criquet, S.; Ziarelli, F.; Buée, M.; Plassard, C.; Taudiere, A.; Garbaye, J.; et al. Into the functional ecology of ectomycorrhizal communities: Environmental filtering of enzymatic activities. *J. Ecol.* **2014**, *104*, 1585–1598.
63. Cullings, K.; Ishkhanova, G.; Henson, J. Defoliation effects on enzyme activities of the ectomycorrhizal fungus *Suillus granulatus* in a *Pinus contorta* (lodgepole pine) stand in Yellowstone National Park. *Oecologia* **2008**, *158*, 77–83. [CrossRef] [PubMed]
64. Kennedy, P.G.; Higgins, L.M.; Rogers, R.H.; Weber, M.G. Colonization-competition tradeoffs as a mechanism driving successional dynamics in ectomycorrhizal fungal communities. *PLoS ONE* **2011**, *9*, e25126. [CrossRef] [PubMed]
65. Pickles, B.J.; Genney, D.R.; Anderson, I.C.; Alexander, I.J. Spatial analysis of ectomycorrhizal fungi reveals that root tip communities are structured by competitive interactions. *Mol. Ecol.* **2012**, *21*, 5110–5123. [CrossRef] [PubMed]
66. Clemmensen, K.E.; Finlay, R.D.; Dahlberg, A.; Stenlid, J.; Wardle, D.A.; Lindahl, B.D. Carbon sequestration is related to mycorrhizal fungal community shifts during long-term succession in boreal forests. *New Phytol.* **2015**, *205*, 1525–1536. [CrossRef] [PubMed]
67. Langley, J.A.; Hungate, B.A. Mycorrhizal controls on belowground litter quality. *Ecology* **2003**, *84*, 2302–2312. [CrossRef]
68. Fernandez, C.W.; Koide, R.T. Initial melanin and nitrogen concentrations control the decomposition of ectomycorrhizal fungal litter. *Soil Biol. Biochem.* **2014**, *77*, 150–157. [CrossRef]
69. Schreiner, K.M.; Blair, N.E.; Levinson, W.; Egerton-Warburton, L.M. Contribution of Fungal Macromolecules to Soil Carbon Sequestration. In *Progress in Soil Science*; Hartemink, A.E., McSweeney, K., Eds.; Springer International Publishing: Cham, Switzerland, 2014; pp. 155–161.

 *forests*

Article

Rapid Shifts in Soil Nutrients and Decomposition Enzyme Activity in Early Succession Following Forest Fire

Joseph E. Knelman [1,*], Emily B. Graham [2], Scott Ferrenberg [3], Aurélien Lecoeuvre [4], Amanda Labrado [5], John L. Darcy [6], Diana R. Nemergut [7,†] and Steven K. Schmidt [6]

[1] Institute of Arctic and Alpine Research, University of Colorado Boulder, 450 UCB, Boulder, CO 80309, USA

[2] Biological Sciences Division, Pacific Northwest National Laboratory, P.O. Box 999, Richland, WA 99352, USA; emily.graham@pnnl.gov

[3] Department of Biology, New Mexico State University, MSC 3AF, P.O. Box 30001, Las Cruces, NM 88003, USA; ferrenbe@nmsu.edu

[4] Université de Bretagne Occidentale, UFR Sciences et Techniques, 6 Avenue Victor Le Gorgeu, 29238 Brest, France; aurelienlecoeuvre@gmail.com

[5] Department of Geological Sciences, The University of Texas at El Paso, 500 W University, El Paso, TX 79902, USA; allabrado719@gmail.com

[6] Department of Ecology and Evolutionary Biology, University of Colorado at Boulder, 334 UCB, Boulder, CO 80309, USA; jack.darcy@colorado.edu (J.L.D.); steve.schmidt@colorado.edu (S.K.S.)

[7] Biology Department, Duke University, 125 Science Drive, Durham, NC 27708, USA

[*] Correspondence: joseph.knelman@colorado.edu; Tel.: +1-303-492-8981; Fax: +1-303-492-6388

[†] Deceased on 31 December 2015.

Received: 3 August 2017; Accepted: 13 September 2017; Published: 15 September 2017

Abstract: While past research has studied forest succession on decadal timescales, ecosystem responses to rapid shifts in nutrient dynamics within the first months to years of succession after fire (e.g., carbon (C) burn-off, a pulse in inorganic nitrogen (N), accumulation of organic matter, etc.) have been less well documented. This work reveals how rapid shifts in nutrient availability associated with fire disturbance may drive changes in soil enzyme activity on short timescales in forest secondary succession. In this study, we evaluate soil chemistry and decomposition extracellular enzyme activity (EEA) across time to determine whether rapid shifts in nutrient availability (1–29 months after fire) might control microbial enzyme activity. We found that, with advancing succession, soil nutrients correlate with C-targeting β-1,4-glucosidase (BG) EEA four months after the fire, and with N-targeting β-1,4-*N*-acetylglucosaminidase (NAG) EEA at 29 months after the fire, indicating shifting nutrient limitation and decomposition dynamics. We also observed increases in BG:NAG ratios over 29 months in these recently burned soils, suggesting relative increases in microbial activity around C-cycling and C-acquisition. These successional dynamics were unique from seasonal changes we observed in unburned, forested reference soils. Our work demonstrates how EEA may shift even within the first months to years of ecosystem succession alongside common patterns of post-fire nutrient availability. Thus, this work emphasizes that nutrient dynamics in the earliest stages of forest secondary succession are important for understanding rates of C and N cycling and ecosystem development.

Keywords: carbon; decomposition; disturbance; ecosystem process; extracellular enzymes; exoenzymes; forest fire; nitrogen; soil enzymes; succession

---

1. Introduction

Global change pressures have increased the prevalence of forest fires in western North America [1–3]. Therefore, a better understanding of the connection between resulting perturbations in environmental factors and ecosystem processes, such as decomposition, will be vital to modeling ecosystem responses in the wake of such disturbance [4–7]. Microbial production of extracellular enzymes (EEA) involved in decomposition is regulated by quantity and quality of substrate that can change strongly after forest fires, although other factors such as moisture and pH are known to impact EEA as well [8–10]. In particular, fire disturbances dramatically alter soil pH, water holding capacity, and carbon (C) and nitrogen (N) pools [11,12], which may all continue to change through succession. These factors may thus affect microbial investment in both C- and N-targeting decomposition enzymes across short time scales of ecosystem recovery (e.g., months).

Past studies in post-fire forest soils have evaluated changes in edaphic properties, microbial communities, and related EEA over decadal time scales [13,14], while shorter-term successional dynamics are less well characterized. However, research has more recently shown the importance of changes in microbial function across short timescales of succession [15–17]. Work in similar forest ecosystems relating to mountain pine beetle kill has shown not only long-term effects of such disturbance, but also immediate, short-timeframe impacts on microbial function, such as respiration [17]. Altogether, this body of work has demonstrated that even when succession is considered over just months to years after a disturbance, shifts in nutrient pools—such as in ammonium (NH_4^+) and C availability—can have strong effects on microbial function [16–18]. Indeed, immediately after fires, burning of soil organic matter leads to alteration of soil C pools; fires can both burn-off C and alter its chemistry [12,19]. For N pools, a pulse of inorganic N occurs immediately after severe burns, which may either be rapidly exported from the ecosystem or persist into the first year following the fire [11,12]. Thus, even on a timescale of months to years after disturbance, these rapid and profound shifts in nutrient pools may influence microbial processes, such as the production of extracellular enzymes, which are central to nutrient cycling and ecosystem dynamics.

Here, we chose to examine early succession on a timescale of months after a high-severity forest fire to understand how soils and microbial enzyme production may change during this time period [20–23], with putative implications for the trajectory of ecosystem development [24,25]. Importantly, we focus on high-severity wildfires that are increasingly prevalent in montane forests of the U.S. Intermountain West [2] and can elicit particular responses in soil edaphic properties and microbial communities that are different from lower-severity burns; studies have shown that burn severity can differentially influence soil nutrient pools, C chemistry transformations, soil physical properties, and microbial community composition (for example [6,18,26]). While past work has evaluated shifts in EEA [14] and environmental controls on enzyme potential across secondary succession in general [27], our study examines how relationships between edaphic properties and decomposition enzyme activity may change within the first years of succession following a severe forest fire. We contrast these relationships between the fire-disturbed (burned) soils and undisturbed (reference) forest soils to highlight the effect of fires versus a control, as well as elucidate successional vs. seasonal patterns.

Specifically, we assessed edaphic properties, nutrient pools, and microbial enzyme activities relating to C and N cycling of soils across three time points in the initial stages of succession spanning 29 months after a major forest fire. We characterized the soil C and N resource environment and β-1,4-glucosidase (BG) and β-*N*-acetylglucosaminidase (NAG) decomposition enzymes that target these resources [9]. Across a global scale, BG and NAG both show strong positive relationships to soil organic matter, NAG shows strong negative relationship with pH, and BG is weakly related to mean annual precipitation [9]. As such, we assessed the activities of these enzymes alongside nutrient pools, pH, and moisture. Enzyme potential and soil nutrient pools are well suited for examining successional patterns as they may broadly persist over inter-seasonal time scales [28] and thus reflect successional patterns beyond seasonal variability.

We hypothesized that on month to year scales of forest secondary succession, changes in soil enzyme activity occur given rapid alterations to edaphic properties after fire disturbance. We further hypothesized that correlations between nutrient pools and EEA would vary between burned and reference soils given the different roles of nutrient limitation (such as stronger C limitations in burned soils); and, that the strength of correlations between BG and NAG with nutrient pools would change over time to reflect common post-fire shifts in nutrient availability in burned soils. Accordingly, for burn soils we hypothesized that enzyme dynamics would reflect rapid shifts in nutrients that occur after fires: initially low C and high N pools would drive correlations between nutrient pools and BG, whereas with accumulating C and reductions in N over time, correlations between nutrient pools and NAG activity would become more prominent [29,30].

2. Methods

2.1. Site and Soil Properties

Samples were collected in the Fourmile Canyon, Boulder County, CO, USA. Samples were collected at 1 month (October 2010), 4 months (January 2011), and 29 months (June 2013) after a major, high-severity wildfire, which ignited on 6 September 2010. At all sample times, replicate samples were taken from both undisturbed (reference) forest soils and adjacent fire-disturbed (burned) soils. These sites were ~300 m apart and sampling areas fell within 650 m^2 landscapes at similar slope-aspect (northeastern facing mountainside slope) and elevation (~2100–2300 meters above sea level) within the extent of the Fourmile Fire (latitude: 40.036153, longitude: −105.400537) (Figure 1). Sample areas were free of tree mortality from bark beetles and fungi. Prior to the burn, this contiguous study area had similar soil conditions across the landscape given consistent topographical and vegetative features (Figure 1). Metamorphic and igneous parent material has resulted in coarse, poorly developed, sandy soils in Fourmile Canyon [31]. Soils are stony sandy loams of the Fern Cliff-Allens Park-Rock outcrop complex as per National Cooperative Soil Survey (NCSS) classification. Annual precipitation averages 475 mm and occurs primarily as snow in winter/spring [31]. Both reference and burned sites were dominated by *Pinus ponderosa* var. *scopulorum* and *Pseudotsuga menziesii* var. *glauca*; details of location vegetation and fire history as well as the Fourmile Canyon Fire dynamics have been previously described [31,32]. Photos and maps of the fire's extent and response to/control of the Fourmile Canyon Fire have been made available by the U.S. Department of Agriculture [31]. Trees in the burned site were severely burned (completely charred) and dead. Unlike the 1- and 4-month soils that were void of vegetation, 29-month soils were revegetated with understory herbaceous plants by seeding, dominated by sterile wheat.

Ten replicates for both burn and forested undisturbed soils were collected at each time point. Sampling locations within each treatment were 1 m from the base of a tree (burned or alive, respectively), and at least 3 m but no more than 25 m between individual trees used for sampling. At each sample location three 130.5 cm^3 soil cores of mineral soil at a depth of 5 cm were taken and bulked to constitute a single replicate. We avoided/removed belowground plant material. In reference site samples, the organic layer was removed prior to sampling; no organic layer was present in burned soils. One-month soils included an ash layer of <0.5 cm.

Within 2 h of sampling, soils were transported to the lab and then sieved through a 2 mm mesh, subsampled, and stored in a −70 °C freezer for molecular analysis or refrigerated at 4 °C for soil chemistry and enzyme assays. All samples were processed according to the methods enumerated in Ferrenberg et al. (2013), from which 1- (October 2010) and 4-month (January 2011) samples from the burned site, and 4-month samples from an undisturbed, unburned, forested site were also used. Soil moisture, pH, total dissolved nitrogen (TDN), extractable, non-purgeable, organic carbon (NPOC), and ammonium (NH$_4^+$) were evaluated.

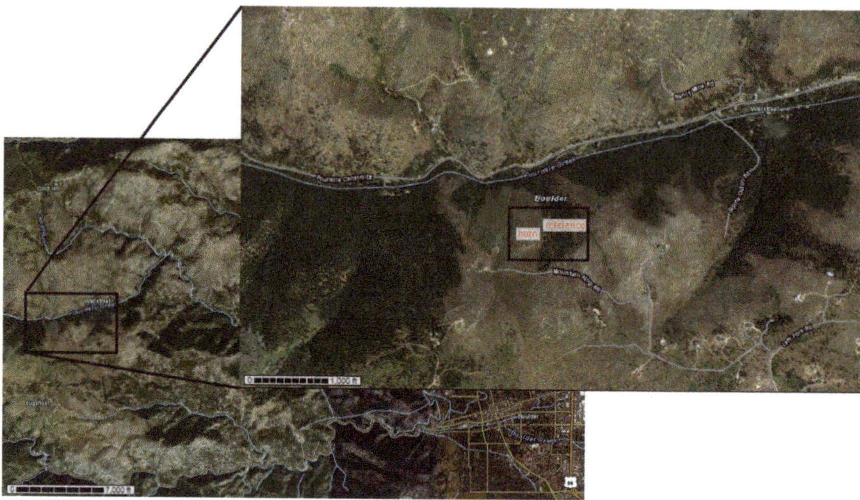

Figure 1. Maps of Boulder and Fourmile Canyon and surrounding region, with sample area demarcated in inset map. Maps are from the U.S. Department of Agriculture Natural Resources Conservation Service.

A subsample of each soil was dried at 100 °C for 48 h to determine gravimetric soil moisture; subsequent edaphic properties were calculated on a dry weight basis. Dried soils of all samples were ground and 50 mg were packed into tin capsules for %C and %N analysis using a Thermo Finnigan EA 1112 Series Flash Elemental Analyzer; (Thermo Fisher Scientific, Inc., Waltham, MA, USA) [33].

Immediately following collection, 8 g of soil were extracted for 1 h in 40 mL of 0.5 M K_2SO_4 and filtered with Whatman No. 1 paper (Whatman Incorporated, Florham Park, NJ, USA). Extract filtrate was frozen until analysis of NH_4^+, TDN, and NPOC. Filtrates were analyzed for NH_4^+ on a BioTek Synergy 2 Multidetection Microplate Reader (BioTek, Winooski, VT, USA) and TDN/NPOC, were measured on a Shimadzu TOC-V CSN Total Organic Carbon Analyzer (Shimadzu TOCvcpn, Kyoto, Japan). TDN, NPOC, and NH_4^+ analysis was completed for all 4 and 29 month soils. Soil pH was measured on soil slurries with a ratio of 2 mg dry soil: 4 mL water, which were shaken at 250 rpm for one hour and allowed to equilibrate for an hour before measuring.

2.2. Enzyme Analysis

Enzyme activities for β-1,4-glucosidase and β-1,4-*N*-acetylglucosaminidase were evaluated to assess microbial investment in C and N acquisition, the cycling of these nutrients, and connections with edaphic properties. BG and NAG enzymes are useful indicators of C and N cycling as they are produced across a wide variety of fungi and bacteria and importantly have been used widely in past research to assess microbial investment in C vs. N acquisition and the limiting nature of these nutrients in post-fire forest ecosystems [9,18,30]. While all ten replicates were used from 29-month samples, due to limited availability of samples, eight replicates of reference forested soils (4 months) and seven replicates of burned soils (1 and 4 months) were included for enzyme analysis. Enzyme activity was measured via fluorometric microplate methods [34,35]. The methods of Weintraub et al. [34,35] were used based on a 96-well assay plate method with 1 M sodium acetate buffer titrated to a pH of 7.0, and 4-methylumbelliferone standards. ~1 g of refrigerated soil was used from each sample [36]. Each sample (every experimental replicate) was run with 16 analytical replicates, quench corrections, standards, and negative controls for each enzyme assay. Fluorescence was measured using a microplate reader (Thermo Labsystems, Franklin, MA, USA) at 365 nm excitation and 460 nm emission to calculate nmol activity h^{-1} g $soil^{-1}$.

2.3. Statistical Analysis

The *pgirmess* package in the R statistical environment [37] was used to evaluate changes in edaphic properties within reference and burn soils across the various time points using Kruskal–Wallis contrasts. Enzyme activity was also analyzed as a BG: NAG ratio and tested for statistical differences across time within both burned and reference forest soils. Pearson product moment correlations were calculated between environmental factors of total C, total N, pH, C:N ratio, and percent moisture and BG/NAG activity both in burned and reference plot samples across all time points. Data were checked for normality and if nonconforming were transformed to achieve normality before correlation analysis.

2.4. Data Availability

All metadata have been made available at figshare [38].

3. Results

3.1. Extracellular Enzyme Activities

In burned soils, BG activity was significantly higher in 29-month soils than one- and four-month soils (Table 1), denoting a trend for increasing activity through time, becoming more comparable to reference soil activity levels. In contrast, NAG activity showed significant declines from 4- to 29-month soils. BG:NAG ratios exhibited a strong partitioning between the 1-/4-month and 29-month time points (Table 1). For instance, in burned samples, 29-month soils had significantly higher BG:NAG ratios than one- and four-month soils.

BG activity at all times was higher in reference soils than burned soils. While BG showed significantly higher activity in 29-month soils than four-month soils, no differences in NAG activity or BG:NAG ratios over time were observed in reference soils (Table 1).

3.2. Soil Properties

Burned soils showed patterns of change over time in ammonium (NH_4^+), total dissolved nitrogen (TDN), and percent moisture (Table 1). Significant decreases in NH_4^+ and TDN were observed between 4- and 29-month burned soils. Moisture declined from 1-month to 29-month time point; soil moisture at one month was significantly higher than at 29 months in burned soils. No significant changes were observed in total C and N pools as measured via %N, %C, or C:N ratio across any of the time points in burned soils.

Unburned reference soils showed declines in soil moisture over time with 4- and 29-month soils have significantly lower soil moisture than one-month soils. pH showed significant differences month to month (Table 1).

3.3. Soil Properties and Extracellular Enzyme Activity

In burned soils, EEA was uncorrelated to edaphic factors initially (i.e., at one month post-fire) but began to strongly relate to nutrient pools (%C and %N) at four months (r of >0.8) and onward (Table 2). These burned soils showed strong correlations between BG (but not NAG) and %C and %N. In contrast, the reference plots at this time point showed correlations of both NAG and BG with edaphic properties including C and N pools. By 29 months, both BG and NAG correlated with C and N pools, while no correlations were observed in reference plots (Table 2). Taken together, these analyses demonstrate that BG activity in burned soils correlated with soil nutrient pools during the 4–29-month post-fire interval, while NAG correlated with these same factors later in successional time only (e.g., at 29 months) (Table 2). In reference soils, both BG and NAG correlated with nutrient pools at four months and showed no correlation at 1- and 29-month time points.

Table 1. Mean (standard deviation) edaphic properties for reference/post-fire successional samples.

Category	pH	% Moisture	% N	% C	NH4 (mg/kg Soil)	NPOC (mg/g Soil)	TDN (mg/g Soil)	BG (nmol Activity/h/g Soil)	NAG (nmol Activity/h/g Soil)	BG:NAG Ratio
BURN										
1-month	7.21 (0.30) [AB]	9.93 (4.55) [BA]	0.14 (0.04)	2.43 (0.95)	N/A	N/A	N/A	53.94 (12.02) [C]	78.53 (31.46) [AB]	0.74 (0.23) [B]
4-months	8.08 (0.41) [A]	8.55 (2.36) [ABC]	0.11 (0.04)	2.33 (1.07)	48.94 (14.95) [A]	0.30 (0.16)	0.06 (0.02) [A]	76.38 (27.67) [BC]	107.54 (43.07) [A]	0.75 (0.18) [B]
29-months	7.00 (0.28) [B]	2.16 (0.80) [C]	0.12 (0.03)	2.48 (0.66)	2.30 (1.33) [B]	0.31 (0.30)	0.02 (0.02) [B]	127.00 (27.57) [A]	44.41 (14.61) [B]	3.01 (0.71) [A]
REFERENCE										
1-month	6.44 (0.67) [B]	19.74 (10.17) [A]	0.27 (0.08)	6.21 (2.01)	N/A	N/A	N/A	N/A	N/A	N/A
4-months	7.08 (0.46) [A]	6.35 (3.50) [C]	0.24 (0.17)	6.59 (5.08)	2.39 (1.40)	0.26 (0.18)	0.02 (0.01)	147.83 (54.97) [B]	136.29 (63.12)	1.17 (0.38)
29-months	6.63 (0.34) [AB]	8.40 (6.12) [BC]	0.24 (0.08)	5.20 (2.19)	1.24 (0.60)	0.09 (0.04)	0.01 (0.004)	254.41 (86.14) [A]	251.77 (116.38)	1.13 (0.38)

Letters denote significant differences across timepoints ($p < 0.05$) as per Kruskal-Wallis contrasts within Burn/Reference categories.

Table 2. Correlations between β-1,4-glucosidase (BG) and β-1,4-*N*-acetylglucosaminidase (NAG) enzyme activity and edaphic properties. Significant ($p < 0.05$) correlations (Pearson's r) shown for burned and reference soils across all time points.

		BURNED PLOTS		REFERENCE PLOTS	
Time	**Factors**	**BG**	**NAG**	**BG**	**NAG**
	pH	NS	NS	N/A	N/A
	moisture	NS	NS	N/A	N/A
1-month post-fire October	C	NS	NS	N/A	N/A
	N	NS	NS	N/A	N/A
	C:N	NS	NS	N/A	N/A
	pH	NS	NS	NS	NS
	moisture	NS	NS	0.9	0.76
4-months post-fire January	C	0.83	NS	0.81	0.79
	N	0.9	NS	0.78	0.72
	C:N	NS	NS	0.77	NS
	pH	NS	NS	NS	NS
	moisture	NS	NS	NS	NS
29-months post-fire June	C	0.69	NS	NS	NS
	N	NS	NS	NS	NS
	C:N	NS	0.69	NS	NS
	pH	NS	NS	NS	NS
	moisture	0.64	NS	NS	NS
33-months post-fire October	C	NS	NS	0.84	0.84
	N	NS	0.66	0.96	0.86
	C:N	NS	0.74	NS	NS

NS = not significant N/A = not available.

4. Discussion

Changes in edaphic properties and EEA of post-fire landscapes have been shown to occur across successional stages at decadal timescales [13,14,30,39,40]. Strikingly, we found that microbial EEA related to C and N acquisition varied significantly over a relatively short time span of 29 months of succession. Our results indicate that even within three years of succession [23] enzyme activity changes alongside rapid shifts in nutrient availability that are characteristic of post-fire succession. While microbes in early succession may be co-limited by both C and other macronutrients such as N [41–43], increasing BG:NAG ratios observed within the first 29 months of post-fire forest succession may reflect increasing C availability (e.g., revegetation) and a relative increase in microbial investment in C acquisition. Reference soils, however, showed no significant changes over time in BG:NAG ratios. Our work is consistent with past research in post high-severity forest fire soils that shows BG:NAG ratios of 2–3 at just over a year into succession, while lower disturbance environments displayed BG:NAG ratio around 1–1.5 [18]. In total, the observed shift demonstrates that within three years of succession EEA activity is responsive to the unique soil nutrient environment of burned soils and shows distinct dynamics from reference forest soils.

We more directly examined the relationship between edaphic properties and EEA within each stage of the burned landscape in contrast to corresponding reference soils. Within these stages of secondary succession, we observed a shift from correlations between only BG EEA and nutrient pools to correlations between NAG EEA and nutrient pools in the 29-month time point as well (Table 2). Although controls on microbial production of extracellular enzymes may vary, it is well known that the quantity and quality of available substrates can induce and structure the production/activity of both C and N acquiring enzymes [9,10]. The observed correlations indicate that even within the first years of secondary succession, nutrient limitation may control BG activity with eventual shifts toward more prominent connections between nutrient pools and NAG activity. This dynamic may reflect a relative shift from C to N limitation (or relative changes in co-limitation) and is consistent with general patterns in nutrient dynamics across succession in post-fire landscapes [11,12,43–45]. Specifically, research has commonly observed that post-fire landscapes are characteristically low in C and experience a pulse of inorganic N in the form of ammonium and nitrate after severe wildfires, while slightly later successional soils may be more constrained by N with the buildup of soil C [11,12]. Accordingly, we witnessed evidence of a pulse of NH_4^+ and TDN in the four-month post-fire soils and a strong drawdown in these N pools at the 29-month time point (Table 2), consistent with a vast body of literature which notes a pulse of inorganic N immediately after a fire, but drawdowns in this pool on a timescale of months to years [11,12]. While C pools do not show significant increases over time in soils at the scale measured in this study, past work has shown that fires can strongly influence the composition of soil organic matter without significant impact on total stock [26]. For example, fires can alter C chemistry in forest soils, including the humification of C compounds which can influence substrate availability for microbial decomposition [18,19,46]. Changes in C pools over successional time with plant colonization (29-month soils) may also be in terms of composition and quality, not just quantity [27,47,48].

While we acknowledge that seasonality can influence variation in EEA [49], the observation that strong correlations between EEA and soil N and C pools correspond with common post-fire dynamics, such as a drawdown in inorganic N, likely reflects successional dynamics. Additionally, the fact that these observed patterns in EEA of successional soils are different from reference soils shows that such patterns are specific post-burn soil dynamics in the first months after a fire, illustrating EEA responses to geochemistry even within 4–29 months post-disturbance.

While pH and moisture have well-described successional dynamics, such as an immediate increase in pH after fire and decreases over time, or increases in water holding capacity with the buildup of soil organic matter over time [11,12], these factors may also vary on a seasonal basis. In the case of this study, we interpret moisture changes, for example, as largely a seasonal shift. Over 29 months, there is little change in soil organic matter and water holding capacity, and shifts in soil moisture occur in a similar manner in both successional and reference soils. This pattern of change is not unique to successional soils, but rather a seasonal dynamic true of reference forests as well. However, neither soil moisture nor pH correlated with enzyme activity in post-burn successional soils. While future work should seek to address how seasonality versus succession influences these ecosystems in the first months after fire, significant increases in BG:NAG ratios over time and correlations between C and N pools with BG/NAG EEA are different from patterns in reference soils and demonstrate dynamics that are unique to post-burn successional soils within the first three years following a severe wildfire.

Additionally, enzymes are well suited to studying inter-seasonal dynamics as they persist in the soil [28] and are assayed for enzyme potential (at controlled temperature, moisture, and pH) rather than in situ enzyme activity. Enzyme potential assays, such as those completed in the lab, may therefore reflect successional dynamics rather than seasonal ones where variable in situ temperature, pH, and moisture can strongly affect enzymatic activity.

Altogether, our work leads to a conceptual model of patterns in the coupling of nutrients and decomposition enzyme activity on short timescales after fires (Figure 2). Because of characteristic changes in nutrient pools over the first years in post-fire succession, and the role of C and N availability as a control on enzyme production, we propose that the initial limitation in C availability results in a connection between BG activity and the resource environment. Likewise, in subsequent stages where C pools begin to build and N is more limited in availability (though C and N may be co-limiting), NAG activity shows connections with the resource environment (Figure 2). Here, in particular with plant colonization and the accumulation of C, BG:NAG ratios increase, reflecting improved availability of C substrates (Figure 2). It is important to note that these dynamics are envisioned for short timescales within the first years of succession, as N limitation alone across longer timescales can yield declines in BG:NAG ratio [30] and more dramatic variation in other important controls such as pH and soil moisture may also become more influential.

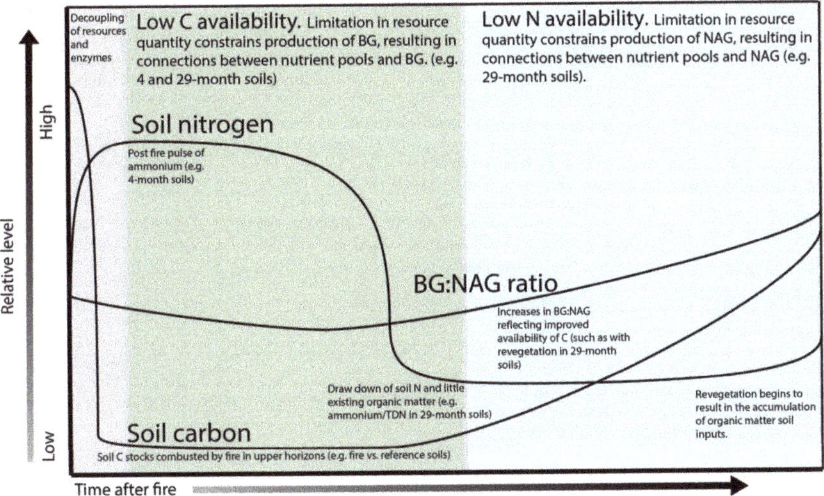

Figure 2. Soil resource and extracellular enzyme dynamics on a short timescale (<3 years) after a forest fire.

5. Conclusions

We found evidence for a connection between rapid shifts in nutrient pools and microbial decomposition enzyme activity in the first several years of secondary succession. We show that within 29 months of post-fire succession relative increases in BG:NAG ratios occur. These shifts are distinct from reference soils, and may represent rapid successional responses to changing nutrient dynamics. Our work demonstrates that soil nutrients first correlate with BG activity (C-targeting) and then correlate additionally with NAG activity (N-targeting) within 29 months of succession. This shift is likely driven by changes in substrate availability and quality as post-fire landscapes first show reductions in C pools, followed by reductions in NH_4^+/TDN pools over the timeframe examined in this study. Built on the empirical findings of this and other studies, our conceptual model suggests when and why we may expect to observe changes in nutrient–enzyme relationships across the initial stages of post-fire succession (Figure 2).

Despite the use of a single site in our research, such study systems and sampling schemes have traditionally been used in the study of ecosystem succession with great success in advancing the field empirically and theoretically [14,44,50,51]. Nonetheless, the research conducted herein represents samples from a single fire disturbance and thus we are limited in our ability to generalize such findings. We also note that scales of disturbance should be explicitly considered in future work and constrain the conclusions of this study, which were based on a high-severity fire. Past work has shown that high- vs. low-severity fires, for example, can modulate ecosystem responses in terms of soil chemistry and EEA [6,18].

We present our conceptual model as a hypothesis for further work (Figure 2). While this work describes shifts in EEA potential and the linkage between nutrients and soil enzyme activity within the first years of succession, future work should more closely examine the possible ecological mechanisms that underlie these patterns, such as how specific changes in microbial communities may be driving the observed differences in biogeochemical potential with EEA. Past work at this site in which bacterial communities were sequenced at each time point showed no correlation between bacterial community structure and EEA in the post-fire successional soils (data not shown); however, fungal communities are dominant drivers of EEA as well and further research may reveal to what extent microbial data can explain variation in soil EEA that is responsible for the cycling of C and N in these ecosystems [5].

Acknowledgments: This work was supported by the National Science Foundation of the USA through grant DEB-1258160 to D.R.N. and S.K.S. We also acknowledge support from the Microbiomes in Transition (MinT) Initiative at Pacific Northwest National Laboratory, operated by Battelle for the U.S. Department of Energy (DE-AC05-76RL01830). We thank Duaba and Sean O'Neill for assistance in the establishment of the field site and Janet Prevéy for botany insights at the field site. We appreciate the expertise of Holly Hughes in analytical chemistry and the comments of two anonymous reviewers on the manuscript. Diana Nemergut—a superbly creative and innovative scientist—guided and worked on this manuscript. Diana continues to impact not only the scientific community and ongoing research but also far wider, diverse communities of people who she forever shines upon with her generous and beautiful life.

Author Contributions: J.E.K., D.R.N., S.K.S., E.B.G. and S.F. conceived and designed the experiments; J.E.K., D.R.N., S.K.S., E.B.G., S.F., and J.D. performed field work; J.E.K., E.B.G., S.F., A.L., and A.L. performed laboratory work; J.E.K. analyzed the data; and J.E.K. wrote the paper with the assistance of D.R., S.K.S., E.B.G., and S.F.

References

1. Westerling, A.L. Warming and Earlier Spring Increase Western U.S. Forest Wildfire Activity. *Science* **2006**, *313*, 940–943. [CrossRef] [PubMed]
2. Miller, J.D.; Safford, H.D.; Crimmins, M.; Thode, A.E. Quantitative Evidence for Increasing Forest Fire Severity in the Sierra Nevada and Southern Cascade Mountains, California and Nevada, USA. *Ecosystems* **2009**, *12*, 16–32. [CrossRef]
3. Rocca, M.E.; Brown, P.M.; MacDonald, L.H.; Carrico, C.M. Climate change impacts on fire regimes and key ecosystem services in Rocky Mountain forests. *For. Ecol. Manag.* **2014**, *327*, 290–305. [CrossRef]
4. Graham, E.B.; Wieder, W.R.; Leff, J.W.; Weintraub, S.R.; Townsend, A.R.; Cleveland, C.C.; Philippot, L.; Nemergut, D.R. Do we need to understand microbial communities to predict ecosystem function? A comparison of statistical models of nitrogen cycling processes. *Soil Biol. Biochem.* **2014**, *68*, 279–282. [CrossRef]
5. Graham, E.B.; Knelman, J.E.; Schindlbacher, A.; Siciliano, S.; Breulmann, M.; Yannarell, A.; Beman, J.M.; Abell, G.; Philippot, L.; Prosser, J.; et al. Microbes as Engines of Ecosystem Function: When Does Community Structure Enhance Predictions of Ecosystem Processes? *Terr. Microbiol.* **2016**, *7*, 214. [CrossRef] [PubMed]
6. Holden, S.R.; Rogers, B.M.; Treseder, K.K.; Randerson, J.T. Fire severity influences the response of soil microbes to a boreal forest fire. *Environ. Res. Lett.* **2016**, *11*, 035004. [CrossRef]
7. Buchkowski, R.W.; Bradford, M.A.; Grandy, A.S.; Schmitz, O.J.; Wieder, W.R. Applying population and community ecology theory to advance understanding of belowground biogeochemistry. *Ecol. Lett.* **2017**, *20*, 231–245. [CrossRef] [PubMed]
8. Fierer, N.; Jackson, R.B. The diversity and biogeography of soil bacterial communities. *Proc. Natl. Acad. Sci. USA* **2006**, *103*, 626–631. [CrossRef] [PubMed]
9. Sinsabaugh, R.L.; Lauber, C.L.; Weintraub, M.N.; Ahmed, B.; Allison, S.D.; Crenshaw, C.; Contosta, A.R.; Cusack, D.; Frey, S.; Gallo, M.E.; et al. Stoichiometry of soil enzyme activity at global scale. *Ecol. Lett.* **2008**, *11*, 1252–1264. [CrossRef] [PubMed]
10. Burns, R.G.; DeForest, J.L.; Marxsen, J.; Sinsabaugh, R.L.; Stromberger, M.E.; Wallenstein, M.D.; Weintraub, M.N.; Zoppini, A. Soil enzymes in a changing environment: Current knowledge and future directions. *Soil Biol. Biochem.* **2013**, *58*, 216–234. [CrossRef]
11. Neary, D.G.; Klopatek, C.C.; DeBano, L.F.; Ffolliott, P.F. Fire effects on belowground sustainability: A review and synthesis. *For. Ecol. Manag.* **1999**, *122*, 51–71. [CrossRef]
12. Certini, G. Effects of fire on properties of forest soils: A review. *Oecologia* **2005**, *143*, 1–10. [CrossRef] [PubMed]
13. Treseder, K.K.; Mack, M.C.; Cross, A. Relationships among fires, fungi, and soil dynamics in alaskan boreal forests. *Ecol. Appl.* **2004**, *14*, 1826–1838. [CrossRef]
14. Holden, S.R.; Gutierrez, A.; Treseder, K.K. Changes in Soil Fungal Communities, Extracellular Enzyme Activities, and Litter Decomposition across a Fire Chronosequence in Alaskan Boreal Forests. *Ecosystems* **2012**, *16*, 34–46. [CrossRef]
15. Nemergut, D.R.; Anderson, S.P.; Cleveland, C.C.; Martin, A.P.; Miller, A.E.; Seimon, A.; Schmidt, S.K. Microbial community succession in an unvegetated, recently deglaciated soil. *Microb. Ecol.* **2007**, *53*, 110–122. [CrossRef] [PubMed]

16. Schmidt, S.K.; Reed, S.C.; Nemergut, D.R.; Grandy, A.S.; Cleveland, C.C.; Weintraub, M.N.; Hill, A.W.; Costello, E.K.; Meyer, A.F.; Neff, J.C.; et al. The earliest stages of ecosystem succession in high-elevation (5000 m above sea level), recently deglaciated soils. *Proc. R. Soc. B* **2008**, *275*, 2793–2802. [CrossRef] [PubMed]

17. Moore, D.J.P.; Trahan, N.A.; Wilkes, P.; Quaife, T.; Stephens, B.B.; Elder, K.; Desai, A.R.; Negron, J.; Monson, R.K. Persistent reduced ecosystem respiration after insect disturbance in high elevation forests. *Ecol. Lett.* **2013**, *16*, 731–737. [CrossRef] [PubMed]

18. Knelman, J.E.; Graham, E.B.; Trahan, N.A.; Schmidt, S.K.; Nemergut, D.R. Fire severity shapes plant colonization effects on bacterial community structure, microbial biomass, and soil enzyme activity in secondary succession of a burned forest. *Soil Biol. Biochem.* **2015**, *90*, 161–168. [CrossRef]

19. Almendros, G.; González-Vila, F.J. Fire-induced transformation of soil organic matter from an oak forest. An experimental approach to the effects of fire on humic substances. *Soil Sci.* **1990**, *149*, 158–168. [CrossRef]

20. Boerner, R.E.J.; Giai, C.; Huang, J.; Miesel, J.R. Initial effects of fire and mechanical thinning on soil enzyme activity and nitrogen transformations in eight North American forest ecosystems. *Soil Biol. Biochem.* **2008**, *40*, 3076–3085. [CrossRef]

21. Docherty, K.M.; Balser, T.C.; Bohannan, B.J.M.; Gutknecht, J.L.M. Soil microbial responses to fire and interacting global change factors in a California annual grassland. *Biogeochemistry* **2011**, *109*, 63–83. [CrossRef]

22. Ferrenberg, S.; O'Neill, S.P.; Knelman, J.E.; Todd, B.; Duggan, S.; Bradley, D.; Robinson, T.; Schmidt, S.K.; Townsend, A.R.; Williams, M.W.; et al. Changes in assembly processes in soil bacterial communities following a wildfire disturbance. *ISME J.* **2013**, *7*, 1102–1111. [CrossRef] [PubMed]

23. López-Poma, R.; Bautista, S. Plant regeneration functional groups modulate the response to fire of soil enzyme activities in a Mediterranean shrubland. *Soil Biol. Biochem.* **2014**, *79*, 5–13. [CrossRef]

24. Nemergut, D.R.; Schmidt, S.K.; Fukami, T.; O'Neill, S.P.; Bilinski, T.M.; Stanish, L.F.; Knelman, J.E.; Darcy, J.L.; Lynch, R.C.; Wickey, P.; et al. Patterns and Processes of Microbial Community Assembly. *Microbiol. Mol. Biol. Rev.* **2013**, *77*, 342–356. [CrossRef] [PubMed]

25. Knelman, J.E.; Nemergut, D.R. Changes in community assembly may shift the relationship between biodiversity and ecosystem function. *Front. Microbiol.* **2014**, *5*. [CrossRef] [PubMed]

26. Miesel, J.R.; Hockaday, W.C.; Kolka, R.K.; Townsend, P.A. Soil organic matter composition and quality across fire severity gradients in coniferous and deciduous forests of the southern boreal region. *J. Geophys. Res. Biogeosci.* **2015**, *120*, 1124–1141. [CrossRef]

27. Cline, L.C.; Zak, D.R. Soil Microbial Communities are Shaped by Plant-Driven Changes in Resource Availability During Secondary Succession. *Ecology* **2015**, *96*, 3374–3385. [CrossRef] [PubMed]

28. Allison, S.D.; Gartner, T.B.; Holland, K.; Weintraub, M.N.; Sinsabaugh, R.L. Soil Enzymes: Linking Proteomics and Ecological Process. In *Manual of Environmental Microbiology*; ASM Press: Washington, DC, USA, 2007; pp. 704–711.

29. Sinsabaugh, R.L.; Hill, B.H.; Follstad Shah, J.J. Ecoenzymatic stoichiometry of microbial organic nutrient acquisition in soil and sediment. *Nature* **2009**, *462*, 795–798. [CrossRef] [PubMed]

30. Gartner, T.B.; Treseder, K.K.; Malcolm, G.M.; Sinsabaugh, R.L. Extracellular enzyme activity in the mycorrhizospheres of a boreal fire chronosequence. *Pedobiologia* **2012**, *55*, 121–127. [CrossRef]

31. Graham, R.; Finney, M.; McHugh, C.; Cohen, J.; Calkin, D.; Stratton, R.; Bradshaw, L.; Nikolov, N. Fourmile Canyon Fire Findings. *Gen. Tech. Rep.* **2012**. [CrossRef]

32. Veblen, T.T.; Kitzberger, T.; Donnegan, J. Climatic and human influences on fire regimes in ponderosa pine forests in the colorado front range. *Ecol. Appl.* **2000**, *10*, 1178–1195. [CrossRef]

33. Matejovic, I. Determination of carbon and nitrogen in samples of various soils by the dry combustion. *Commun. Soil Sci. Plant Anal.* **1997**, *28*, 1499–1511. [CrossRef]

34. Sinsabaugh, R.; Carreiro, M.; Repert, D. Allocation of extracellular enzymatic activity in relation to litter composition, N deposition, and mass loss. *Biogeochemistry* **2002**, *60*, 1–24. [CrossRef]

35. Weintraub, S.R.; Wieder, W.R.; Cleveland, C.C.; Townsend, A.R. Organic matter inputs shift soil enzyme activity and allocation patterns in a wet tropical forest. *Biogeochemistry* **2012**, *114*, 313–326. [CrossRef]

36. Bueno de Mesquita, C.P.; Knelman, J.E.; King, A.J.; Farrer, E.C.; Porazinska, D.L.; Schmidt, S.K.; Suding, K.N. Plant colonization of moss-dominated soils in the alpine: Microbial and biogeochemical implications. *Soil Biol. Biochem.* **2017**, *111*, 135–142. [CrossRef]

37. R Development Core Team. *R: A Language and Environment for Statistical Computing*; R Foundation for Statistical Computing: Vienna, Austria, 2013.

38. Figshare. Available online: https://doi.org/10.6084/m9.figshare.1556158.v1 (accessed on 14 September 2017).
39. Allison, S.D.; Gartner, T.B.; Mack, M.C.; McGuire, K.; Treseder, K. Nitrogen alters carbon dynamics during early succession in boreal forest. *Soil Biol. Biochem.* **2010**, *42*, 1157–1164. [CrossRef]
40. Dooley, S.R.; Treseder, K.K. The effect of fire on microbial biomass: A meta-analysis of field studies. *Biogeochemistry* **2012**, *109*, 49–61. [CrossRef]
41. Göransson, H.; Olde Venterink, H.; Bååth, E. Soil bacterial growth and nutrient limitation along a chronosequence from a glacier forefield. *Soil Biol. Biochem.* **2011**, *43*, 1333–1340. [CrossRef]
42. Yoshitake, S.; Uchida, M.; Koizumi, H.; Nakatsubo, T. Carbon and nitrogen limitation of soil microbial respiration in a High Arctic successional glacier foreland near Ny-Ålesund, Svalbard. *Polar Res.* **2007**, *26*, 22–30. [CrossRef]
43. Knelman, J.E.; Schmidt, S.K.; Lynch, R.C.; Darcy, J.L.; Castle, S.C.; Cleveland, C.C.; Nemergut, D.R. Nutrient Addition Dramatically Accelerates Microbial Community Succession. *PLoS ONE* **2014**, *9*, e102609. [CrossRef] [PubMed]
44. Chapin, F.S.; Walker, L.R.; Fastie, C.L.; Sharman, L.C. Mechanisms of primary succession following deglaciation at Glacier Bay, Alaska. *Ecol. Monogr.* **1994**, *64*, 149–175. [CrossRef]
45. Vitousek, P.M.; Farrington, H. Nutrient limitation and soil development: Experimental test of a biogeochemical theory. *Biogeochemistry* **1997**, *37*, 63–75. [CrossRef]
46. Neff, J.C.; Harden, J.W.; Gleixner, G. Fire effects on soil organic matter content, composition, and nutrients in boreal interior Alaska. *Can. J. For. Res.* **2005**, *35*, 2178–2187. [CrossRef]
47. Knelman, J.E.; Legg, T.M.; O'Neill, S.P.; Washenberger, C.L.; González, A.; Cleveland, C.C.; Nemergut, D.R. Bacterial community structure and function change in association with colonizer plants during early primary succession in a glacier forefield. *Soil Biol. Biochem.* **2012**, *46*, 172–180. [CrossRef]
48. Yuan, X.; Knelman, J.E.; Gasarch, E.; Wang, D.; Nemergut, D.R.; Seastedt, T.R. Plant community and soil chemistry responses to long-term nitrogen inputs drive changes in alpine bacterial communities. *Ecology* **2016**, *97*, 1543–1554. [CrossRef] [PubMed]
49. Bardgett, R.D.; Bowman, W.D.; Kaufmann, R.; Schmidt, S.K. A temporal approach to linking aboveground and belowground ecology. *Trends Ecol. Evol.* **2005**, *20*, 634–641. [CrossRef] [PubMed]
50. Nemergut, D.R.; Knelman, J.E.; Ferrenberg, S.; Bilinski, T.; Melbourne, B.; Jiang, L.; Violle, C.; Darcy, J.L.; Prest, T.; Schmidt, S.K.; et al. Decreases in average bacterial community rRNA operon copy number during succession. *ISME J.* **2015**, *10*, 1147. [CrossRef] [PubMed]
51. Castle, S.C.; Nemergut, D.R.; Grandy, A.S.; Leff, J.W.; Graham, E.B.; Hood, E.; Schmidt, S.K.; Wickings, K.; Cleveland, C.C. Biogeochemical drivers of microbial community convergence across actively retreating glaciers. *Soil Biol. Biochem.* **2016**, *101*, 74–84. [CrossRef]

Article

Effects of Nitrogen Deposition on Soil Dissolved Organic Carbon and Nitrogen in Moso Bamboo Plantations Strongly Depend on Management Practices

Zhaofeng Lei [1,†], **Huanfa Sun** [1,†], **Quan Li** [1], **Junbo Zhang** [1] **and Xinzhang Song** [1,2,*]

[1] State Key Laboratory of Subtropical Silviculture, Zhejiang A&F University, Hangzhou 311300, China; leizhaofeng777@163.com (Z.L.); sunhuanfa_zafu@163.com (H.S.); quanlizjafu@163.com (Q.L.); zhangjunbo12138@163.com (J.Z.)

[2] Tianmu Mountain Forest Ecosystem Research Station, Hangzhou 311300, China

* Correspondence: songxinzhang@gmail.com; Tel.: +86-0571-6374-1816

† These authors contributed equally to this work and should be considered co-first authors.

Received: 12 September 2017; Accepted: 13 November 2017; Published: 17 November 2017

Abstract: Soil dissolved organic carbon (DOC) and nitrogen (DON) play significant roles in forest carbon, nitrogen and nutrient cycling. The objective of the present study was to estimate the effect of management practices and nitrogen (N) deposition on soil DOC and DON in Moso bamboo (*Phyllostachys edulis* (Carrière) J. Houz) plantations. This experiment, conducted for over 36 months, investigated the effects of four N addition levels (30, 60 and 90 kg N ha^{-1} year^{-1}, and the N-free control) and two management practices (conventional management (CM) and intensive management (IM)) on DOC and DON. The results showed that DOC and DON concentrations were the highest in summer. Both intensive management and N deposition independently decreased DOC and DON in spring ($p < 0.05$) but not in winter. However, when combined with IM, N deposition increased DOC and DON in spring and winter ($p < 0.05$). Our results demonstrated that N deposition significantly increased the loss of soil DOC and DON in Moso plantations, and this reduction was strongly affected by IM practices and varied seasonally. Therefore, management practices and seasonal variation should be considered when using ecological models to estimate the effects of N deposition on soil DOC and DON in plantation ecosystems.

Keywords: soil organic carbon; dissolved organic matter; nitrogen addition; *Phyllostachys edulis*

1. Introduction

Dissolved organic matter (DOM) is a mixture of organic molecules of different sizes and structures that can pass through a 0.45 µm sieve and dissolve in water and acidic and alkaline solutions [1,2]. Although DOM accounts for a small proportion of the soil organic matter pool, it plays an important role in microbial growth and metabolism, which regulate soil nutrient loss and affect the decomposition and transformation of soil organic matter [3,4]. As the two important components of DOM, dissolved organic carbon (DOC) and nitrogen (DON) are the two common empirical indices that reflect the quantitative characteristics of DOM [5]. DOC affects the regulation of cation leaching, mineral weathering, soil microbial activity, and anion adsorption and desorption, as well as other soil chemical, physical and biological processes [6]. DOC is a very important and active factor associated with terrestrial and aquatic ecosystems in the geochemical carbon cycle [6,7]. DON plays a dual role in N cycling in terrestrial ecosystems: on the one hand, it can be directly absorbed by plants and therefore shorten the terrestrial nitrogen cycle; on the other hand, however, DON has high mobility and can cause pollution in aquatic ecosystems through surface runoff or leaching [8]. DOC and DON have been recognized as the key components of forest C, N and nutrient cycling [9] and are therefore receiving a great deal of attention by researchers.

With the intensification of human activities, global atmospheric N deposition has increased rapidly, and several global N models predict that the subtropical regions in south central China will be among the areas most severely affected by atmospheric N deposition in the coming decades [10]. N is one of the essential elements for plant growth; thus, N deposition is thought to be beneficial to the plant. However, a few studies have shown that only a small proportion of atmospheric N deposited in forest ecosystems is utilized by plants, and that the major proportion of N is fixed in the soil [11]. Previous studies have shown the effects of N deposition on soil DOC and DON in forests. Frey et al. [12] found that more than half of the ecosystem C storage in a hardwood stand was attributable to an accumulation of soil organic matter, indicating that the soil has been more responsive to N addition than tree growth. Furthermore, they thought N enrichment resulted in a shift in organic matter chemistry and microbial community, thereby impacting DOM. Findlay et al. [13] reported that N deposition induces a great loss of soil DOC. Based on N saturation experiments, Gundersen et al. [14] showed that N input can increase the stability of soil humus and promote bacterial growth, thereby leading to a decrease in soil DON. Tu et al. [15] observed that N deposition significantly reduces soil microbial biomass carbon (MBC) but increases DOC in Sinocalmus bamboo (*Neosinocalamus affinis* (Rendle) Keng f.) plantations. However, the effects of N deposition on soil DOC and DON in Moso bamboo (*Phyllostachys edulis* (Carrière) J. Houz) plantations remain unknown.

Because of their rapid growth rate and high annual regrowth rate after harvesting, Moso bamboo forests are the most important source of non-wood forest products in China; they cover an area of 4.43 million ha and represent 73.7% of the country's bamboo forest area and 84.0% of the global distribution of Moso bamboo [16,17]. Our long-term investigation based on the Eddy covariance method showed that the Moso bamboo plantation ecosystem has a high C uptake capacity and might play an important role in mitigating climate warming [17]. In recent decades, intensive management (IM) has been implemented in more than half of these bamboo plantations to increase economic benefits. IM includes the removal of understory weeds, application of fertilizers, and soil tilling [18]. Typically, conventional management (CM) requires the regular harvest of bamboo stems and shoots, without any of the IM practices mentioned [19]. Thus, IM can affect soil organic carbon (SOC) [20], soil microbial biomass [21], and enzyme activities [22], which may affect the soil DOM. Moso bamboo plantations are mainly distributed in subtropical China, which is suffering severe N deposition of 30–37 kg N ha^{-1} year^{-1} [19]. Our previous study [21] found that IM and N deposition significantly increased soil MBC but decreased bacterial diversity, and the combination of management practices and N deposition had greater effects on soil microbial biomass and diversity than either practice system or N deposition independently, which may impact soil DOC and DON. However, the mechanism of DOM response to these complicated factors and subsequent C and N cycles in Moso bamboo plantations remains unknown.

We conducted a more than three-year-long field experiment in Moso bamboo plantations to test the following three hypotheses: (1) IM practices decrease soil DOC and DON; (2) N deposition decreases soil DOC and DON; and (3) the combined effects of N deposition and management practices on soil DOC and DON are stronger than the effects of each of these factors independently.

2. Materials and Methods

2.1. Study Site

The study site was in Qingshan Town, Lin'an City (30°14′ N, 119°42′ E), Zhejiang Province, China. It has a subtropical monsoon climate, with mean annual precipitation and mean annual temperature of 1420 mm and 15.6 °C, respectively. The area receives an average of approximately 1847 h of sunshine and 230 frost-free days annually.

The CM Moso bamboo forests were originally established in the late 1970s from native evergreen broadleaf forests in sites of similar topography (southwest slope of approximately 6°) and soil type. The soils are named yellow-red soil and classified as Ferrisols derived from granite [19]. IM practices were conducted in half of the CM Moso bamboo forests since 2001. In September of each year, the IM Moso bamboo forests were fertilized and then plowed to a depth of 0.3 m. The application of the nitrate of S-based compound fertilizer (N-P_2O_5-K_2O: 15%-6%-20%, 450 kg ha^{-1}) is equivalent to the annual addition of 67.5 kg N, 11.8 kg P, and 74.7 kg K per hectare [19].

2.2. Experimental Design and N Treatment

Twelve CM plots and 12 IM plots, each with an area of 20 m × 20 m, were established. Detailed information on the experimental design can be found in Song et al. [19]. The local background atmospheric N deposition rate is 30–37 kg N ha^{-1} year^{-1}, with an average NH_4^+:NO_3^- of 1.28 [23]. Therefore, NH_4NO_3 was used as the N source for the low-N (30 kg ha^{-1} year^{-1}) treatment (N30), medium-N (60 kg ha^{-1} year^{-1}) treatment (N60), and high-N (90 kg ha^{-1} year^{-1}) treatment (N90). Three replicate plots for each treatment and the control (N-free) were randomized for each management practice. From January 2013, appropriate quantities of NH_4NO_3 were dissolved in 10 L water and sprayed evenly onto the forest floor of the corresponding plot every month. Each control plot received 10 L of N-free water every month to balance the effects of the water added.

2.3. Soil Sampling and Measurement

The experimental plots were sampled in the spring (30 April), summer (25 July), and winter (29 December) of 2016. The monthly mean air temperatures and precipitation quantities during the study period are shown in Table 1. Five soil cores at a depth of 0–20 cm were randomly collected from each plot and mixed. The samples were kept in an incubator, brought to the laboratory, and then sieved through a 2 mm mesh to remove roots, plant residues and stones. Next, we weighed two samples of 20 g fresh soil each and named them A and B. A was for determining DOC and total dissolved nitrogen (TDN) concentrations, and B was for determining soil moisture content. A was extracted with distilled water (soil:water ratio, 2:1), shaken for 0.5 h (170 rpm) at 25 °C, centrifuged for 20 min at 3500 rpm, and then filtered through a membrane (0.45 µm, Millipore, Xingya Corporation, Shanghai, China) into a plastic bottle [24]. DOC and TDN concentrations were determined using a total organic carbon analyzer (TOC-V$_{CPH}$, Shimadzu Corporation, Kyoto, Japan), and NH_4^+-N and NO_3^--N concentrations were determined using the SmartChem 200 Discrete Analyzer. DON was calculated as the difference between TDN and NH_4^+-N and NO_3^--N concentrations, according to Li et al. [24]. The ppm units of DOC, TDN, NH_4^+-N, and NO_3^--N were converted to mg/kg by the following formula:

$$M = \frac{P \times 50 \times 10}{20 \times (1+S)} \tag{1}$$

where M is the combined value of DOC, TDN, NH_4^+-N and NO_3^--N in mg/kg; P is value of DOC, TDN, NH_4^+-N and NO_3^--N in ppm; 50 is the transformation factor from ppm to mg/kg; 10 is the dilution factor; 20 stands for 20 g of the soil sample; and S is the soil moisture content.

Table 1. Average monthly climatic data of the study site during the experimental periods in 2016.

Months	Total Monthly Precipitation (mm)	Average Monthly Air Temperature (°C)
April	352.5	11.49
July	244.8	19.34
December	67.2	3.46

2.4. Data and Statistical Analyses

One-way analysis of variance (ANOVA) and the least significant difference (LSD) method were used to determine the statistical significance of the differences in DOC and DON concentration among N addition treatments under the same management and season, between the two management practices under the same N addition treatment and season, and among three seasons under the same management and N addition treatment.

Two-way ANOVA was performed to evaluate the combined influence of N deposition and management practices in each season. All data were tested for homogeneity of variance and normality of distribution prior to conducting the ANOVA. The data satisfied the assumption of homogeneity of variance. These analyses were performed using SPSS 22.0 (SPSS Inc. Chicago, IL, USA) and SigmaPlot 12.5 for Windows.

3. Results

3.1. Soil DOC

In the CM plots, N deposition significantly reduced DOC concentration in April and July, but not in December (Figure 1). Moreover, in July, the DOC concentration under N60 treatment was significantly higher than that under N30 or N90. In the IM plots, the effect of N deposition on DOC concentration was similar to that in the CM plots in July (Figure 1). However, this effect was not significant in April and December. Nonetheless, a significant increase was observed under the N90 treatment.

The DOC concentrations in the CM plots were significantly higher than those in the IM plots in April and July, but not in December (Figure 1). Moreover, DOC concentration was significantly higher in July than in April and December under both management practices (Figure 1).

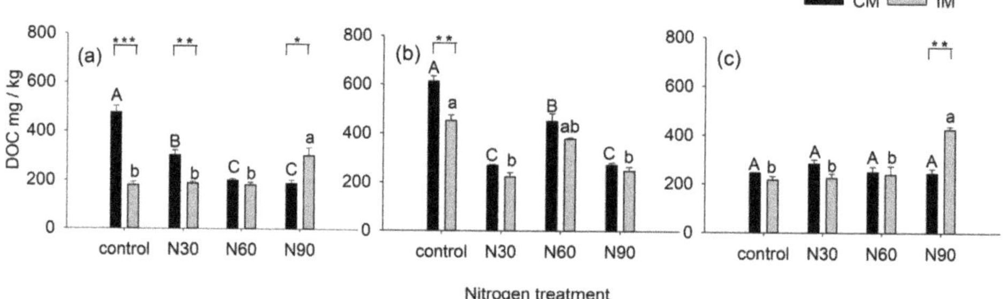

Figure 1. Dissolved organic carbon (DOC) in surface soil (0–20 cm) in different seasons ((**a**) April; (**b**) July; (**c**) December) under different management practices (CM: conventional management; IM: intensive management) and four nitrogen addition treatments (N30: 30 kg N ha^{-1} year^{-1}; N60: 60 kg N ha^{-1} year^{-1}; N90: 90 kg N ha^{-1} year^{-1} and Control: N-free). Vertical bars indicate the standard error of three replicates. Different uppercase letters indicate significant differences among N addition rates under CM treatments ($p < 0.05$). Different lowercase letters indicate significant differences among N addition rates under IM treatments ($p < 0.05$). Asterisks indicate significant differences between CM and IM at the same N addition rate (* $p < 0.05$, ** $p < 0.01$, *** $p < 0.001$).

3.2. Soil DON

In the CM plots, N deposition significantly decreased DON concentrations in April, but not in July and December, except for the N30 treatment in July ($p < 0.05$) (Figure 2). In the IM plots, only the N60 and N90 treatments significantly increased DON concentrations in both April and December, but not in July (Figure 2). The DON concentration was significantly higher in the CM plots than in the

IM plots in April, but not in July and December (Figure 2). Similar to DOC, DON concentration was significantly higher in July than in April and December, under both management practices (Figure 2).

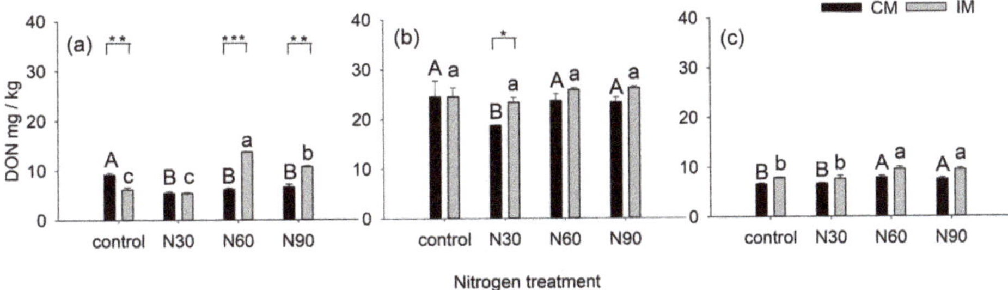

Figure 2. Dissolved organic nitrogen (DON) in surface soil (0–20 cm) in different seasons ((**a**) April; (**b**) July; (**c**) December) under different management practices (CM: conventional management; IM: intensive management) and four nitrogen addition treatments (N30: 30 kg N ha^{-1} year^{-1}; N60: 60 kg N ha^{-1} year^{-1}; N90: 90 kg N ha^{-1} year^{-1} and Control: N-free). Vertical bars indicate the standard error of three replicates. Different uppercase letters indicate significant differences among N addition rates under CM treatments ($p < 0.05$). Different lowercase letters indicate significant differences among N addition rates under IM treatments ($p < 0.05$). Asterisks indicate significant differences between CM and IM at the same N addition rate (* $p < 0.05$, ** $p < 0.01$, *** $p < 0.001$).

3.3. Combined Influence of N Deposition and Management on Soil DOC and DON

The two-way ANOVA showed that N deposition and management practices, independently and in combination, significantly affected DOC and DON in April ($p < 0.001$). In addition, the contribution of the interaction was greater than the independent effects of the two factors (Table 2). In July, N deposition and management practices, independently and in combination, significantly affected DOC ($p < 0.05$), whereas DON was significantly affected by the independent factors only ($p < 0.01$) (Table 2). Moreover, the contribution of separate factors was greater than of their interaction. In December, N deposition significantly affected the DOC only, whereas the management practices significantly affected DON only ($p < 0.05$), and their interaction significantly affected the DOC only (Table 2). The contribution of interaction was greater than that of the two factors separately on DOC only.

Table 2. Two-way ANOVA of the effects of N deposition and management practices on soil dissolved organic carbon (DOC) and nitrogen (DON) at 0–20 cm soil depths in Moso bamboo forests.

Months	Source of Variation/Factors	N Deposition			Management Practices			Interaction		
		F Value	*p* Value	Contribution (%) [a]	*F* Value	*p* Value	Contribution (%)	*F* Value	*p* Value	Contribution (%)
April	DOC	**18.98**	0.0000	22.00	**38.67**	0.0000	14.94	**49.07**	0.0000	56.88
	DON	**58.31**	0.0000	33.81	**74.61**	0.0000	14.42	**83.96**	0.0000	48.68
July	DOC	**66.26**	0.0000	82.92	**14.37**	0.0016	5.99	**3.53**	0.0390	4.42
	DON	**6.82**	0.0036	28.98	**32.99**	0.0000	46.74	0.38	0.7700	1.61
December	DOC	**6.23**	0.0052	26.46	0.36	0.5553	0.51	**11.87**	0.0002	50.38
	DON	2.30	0.1161	19.95	**7.54**	0.0143	21.80	1.39	0.2831	12.02

Significant contribution at $p < 0.05$ or $p < 0.01$ is shown in bold. [a] The contribution (%) is the percentage of overall variance explained by each factor.

4. Discussion

4.1. Effects of Management Practices on Soil DOC and DON

The present study showed that IM significantly decreased DOC and DON concentrations in spring (Figures 1 and 2), which partly supported our first hypothesis: IM practices decrease soil DOC and DON. Zhou et al. [25] showed that the total soil dissolved C at 0–20 cm soil depth in the IM Moso bamboo plantations was lower than that in the CM plots, which is consistent with the results of the present study. It is known that the concentrations of DOC and DON are mainly derived from ground litter, root exudates, soil humus, soil microbial biomass, and rainfall leaching [26]. Wu et al. [27] found that excessive consumption by microbial populations would decrease the DOC and DON in seasons with high temperature and high precipitation, which is consistent with our result. DOC and DON are important carriers of C and N loss in forest soil [28,29]. Furthermore, soil microbial consumption and leaching are the main output pathways of DOC and DON from forest ecosystems [30]. Yang et al. [31] proved that the growth of plant roots and soil microorganisms (represented by MBC) was enhanced by fertilization, increasing the amount of organic compounds (i.e., DOM) released by plant roots and soil microorganisms. Our previous study on this site showed that IM significantly increased soil MBC [21], indicating an increase in DOC consumption, which might greatly contribute to lower DOC concentrations in the IM plots than in the CM plots. Changed nutrient dynamics caused by management practices can also affect DOM concentrations between the native forests and plantations [27]. Long-term fertilization in the IM plots induced the loss of soil organic C and N and greatly decreased the chemical activity of the soil [20]. Soil acidification owing to long-term fertilization in the IM plots was more severe than that in the CM plots, which led to lower soil pH [19]. The decrease in soil pH might increase the adsorption capacity of Fe and/or Al oxides in soil [32], thereby reducing DOC and DON in the IM plots. Vance et al. [33] also reported a similar result. Generally, the DOM concentration is higher in forest topsoil than in cultivated soil, and plowing, weeding and fertilization in the IM plots alter the physical structure of soil, thus increasing the loss of DOC and DON through surface runoff and subsurface flow [34]. This, combined with high precipitation in spring (Figure 1), might lead to more export of DOC and DON from the IM plots and thereby decrease these concentrations more strongly in the IM than in the CM plots. The increased leaching effect on DOC in the rainy season was also observed by Neff and Asner [35]. Therefore, the high precipitation in spring (Table 1) might have washed a large amount of DOM and caused its loss by leaching, which might have contributed to the decline in DOC and DON between CM and IM in April and July.

In summer, DOC and DON concentrations were higher than those in spring and winter. However, IM largely reduced the concentration of DOC but not of DON in summer (Figures 1 and 2). Temperature can affect soil DOM concentration and turnover by controlling microbial biomass [36]. Jiang et al. [37] found that there was twice as much microbial biomass in summer than in spring in Moso bamboo plantations; thus, the high temperature in July (Table 1) might have contributed to the high DOC and DON. Furthermore, compared with spring, summer and winter had lower precipitation (Table 1), thereby inducing a smaller leaching loss of DOC and DON. At this point, the N input from fertilization might have contributed more to the slightly higher DON in the IM plots than that in the CM plots.

In winter, DOC and DON concentrations were low and did not show significant differences between IM and CM (Figures 1 and 2). A possible reason is that the low temperature in winter decreased the positive effects of temperature on DOC and DON. Moreover, the leaching loss of DON induced by plowing in the IM plots might have declined due to low precipitation (Table 1) and might have even been offset by N input from fertilization, which can contribute to slightly higher DON in the IM plots than that in the CM plots.

4.2. Effects of N Deposition on Soil DOC and DON

In the present study, N addition significantly decreased DOC and DON in the CM plots in spring (Figures 1 and 2), which partly supports our second hypothesis: N deposition decreases soil DOC and DON. N saturation experiments showed that exogenous N input can increase the leaching loss of DON [14] and induce a decrease in soil DON. Previous studies have reported that long-term N deposition reduces soil MBC [22,38]. A high degree of N deposition could affect the composition of the microbial community and inhibit the C of microbial degradation, thereby decelerating the decomposition of litter [39,40]. The effect mentioned above contributes to a decline in soil DOC and DON. Notably, N deposition can intensify soil acidification and decrease soil pH [21], which can potentially change the acidity of soil solutions [41–44]. This, in turn, may lead to the decline in DOC and DON in the CM plots.

In summer, N deposition significantly decreased the DOC concentration but did not significantly affect DON, except under the N30 treatment (Figures 1 and 2). In contrast to DOC, DON increased substantially from April to July, which might be attributed to the stronger adsorption of soil to DON than DOC at high temperatures [45]. Moreover, the high-temperature effect might be offset by the negative effect of N deposition on DON. DOC and DON concentrations in winter were not affected by N deposition (Figures 1 and 2). Probably, the low soil microbial activity due to low temperature and precipitation in winter alleviated the effect of N deposition.

4.3. Combined Influence of N Deposition and Management Practices on Soil DOC and DON

When combined with IM, N deposition significantly increased DOC and DON concentrations in spring and winter. This effect was opposite to that of the independent effect of N deposition in the CM plots. This result indicated that intensive management practices may change the direction of the effects of N deposition on DOC and DON (Figures 1 and 2), which did not support our third hypothesis: the combined effects of N deposition and management practices on soil DOC and DON are stronger than the effects of each of these factors independently. Our previous study at this site found that high N deposition decreased the decomposition of leaf litter and fine roots in IM plots [19,23], which could lead to more litter and fine root accumulation on the soil surface and subsurface. This accumulation may alleviate the leaching of DOC through surface runoff and subsurface flow, which partially explains the current finding that high N addition increased DOC in the IM plots. Nevertheless, in summer, the combined effects of N deposition and IM on DOC and DON were similar to the independent effect of N deposition in the CM plots (Figures 1 and 2), which indicated that the effect of interaction of N deposition and management practices on DOC and DON can vary with season.

The two-way ANOVA demonstrated that in spring, the combined influence of N deposition and management practices had greater effects on soil DOC and DON than the effects of each of these factors independently, which supported our third hypothesis (Table 2). Our previous study [21] at this site demonstrated that differences in microbial community structure were primarily due to a combination of N deposition and management practices (57.73%), with management practices alone accounting for 36.26% of the variation and N addition accounting for 21.47%, indicating that the combination of two factors has a stronger impact on DOM than each factor singly, which also supported our present result. In summer, the positive effects of high temperature partially offset the negative effects of N deposition and IM alone on DOC and DON, thus contributing to the low combination of these two factors (Table 2). Our previous study at this site elucidated that N deposition significantly decreased the diversity of soil microorganisms in both CM and IM plots [21], which indicates that the soil DOC and DON may be strongly correlated with microbial diversity in Moso bamboo plantations. Our results suggest, however, that N deposition has a higher contribution than management practices to soil DOC and DON (Table 2), except for DON in summer. Both DOC and DON concentrations showed the same tendency in the control treatment or the combined effects of N deposition and management practices: DOC and DON increased from spring to summer, then decreased in winter. It appears to be the combination of high precipitation and temperature that increased soil adsorption to DOC, especially for DON [45].

5. Conclusions

The present study showed that DOC and DON concentrations were higher in summer than in spring and winter in Moso bamboo plantations. Both IM practices and N deposition independently decreased DOC and DON in spring, but not in winter. When combined with IM, N deposition increased DOC and DON in spring and winter. The effects of N deposition on soil DOC and DON strongly depended on management practices and season, suggesting that management practices and seasonal variation should be taken into account when applying ecological models to estimate the effects of N deposition on DOC and DON in terrestrial ecosystems, also indicating that anthropogenic management practices such as plowing and weeding would partially offset the negative effects of N deposition on DOM in the Moso bamboo plantation ecosystem. The results of the present study provide a new perspective for improving our understanding of the comprehensive effects of N deposition and management practices on C and N cycling in plantations. Our findings also offer a beneficial reference on how to manage plantations under the current background of increasing levels of atmospheric N deposition.

Supplementary Materials: The following are available online at www.mdpi.com/1999-4907/8/11/452/s1, Figure S1: Ammonium nitrogen (NH_4^+-N), Nitrate nitrogen (NO_3^--N) and Total dissolved nitrogen (TDN) in surface soil (0–20 cm) in different seasons ((a), (d), (g): April; (b), (e), (h): July; (c), (f), (i): December) under different management practices (CM: conventional management; IM: intensive management) and four nitrogen addition treatments (N30: 30 kg N ha^{-1} year^{-1}; N60: 60 kg N ha^{-1} year^{-1}; N90: 90 kg N ha^{-1} year^{-1} and Control: N-free). Vertical bars indicate the standard error of three replicates. Different uppercase letters indicate significant differences among N addition rates under CM treatments ($p < 0.05$). Different lowercase letters indicate significant differences among N addition rates under IM treatments ($p < 0.05$). Asterisks indicate significant differences between CM and IM at the same N addition rate (* $p < 0.05$, ** $p < 0.01$, *** $p < 0.001$), Table S1: The initial stand and soil characteristics of the study sites in the Moso bamboo forest (mean $\pm$ SD, $n = 4$).

Acknowledgments: This study was funded by the National Natural Science Foundation of China (Grant Nos. 31470529, 31270517).

Author Contributions: Z.L. and H.S. analyzed the data and drafted the manuscript, and participated in collecting the experiment data. Q.L. and J.Z. were involved in data sampling. X.S. designed this experiment and drafted the manuscript. All authors discussed the results and revised the manuscript.

Conflicts of Interest: The authors declare no conflict of interest.

References

1. Kalbitz, K.; Solinger, S.; Park, J.H.; Michalzik, B.; Matzner, E. Controls on the dynamics of dissolved organic matter in soils: A review. *Soil Sci.* **2000**, *165*, 277–304. [CrossRef]
2. He, W.; Chen, M.; Schlautman, M.A.; Hur, J. Dynamic exchanges between DOM and POM pools in coastal and inland aquatic ecosystems: A review. *Sci. Total Environ.* **2016**, *551–552*, 415–428. [CrossRef] [PubMed]
3. Zhao, J.; Zhang, X.; Yuan, X.; Wang, J. Characteristics and environmental significance of soil dissolved organic matter. *Chin. J. Appl. Ecol.* **2003**, *14*, 126–130.
4. Hopkinson, C.S.; Vallino, J.J. Efficient export of carbon to the deep ocean through dissolved organic matter. *Nature* **2005**, *433*, 142–145. [CrossRef] [PubMed]
5. Xiao, H.; Liu, B.; Yu, Z.; Wan, X.; Sang, C.; Zhou, F.; Huang, Z. Effects of forest practices on soil dissolved organic carbon and nitrogen in surface and deep layers in subtropical region, China. *Chin. J. Appl. Ecol.* **2016**, *27*, 1031–1038.
6. Li, S.; Yu, Y.; He, S. Summary of research on dissolved organic carbon (DOC). *Soil. Environ. Sci.* **2002**, *11*, 422–429.
7. Rosemond, A.D.; Benstead, J.P.; Bumpers, P.M.; Gulis, V.; Kominoski, J.S.; Manning, D.W.; Suberkropp, K.; Wallace, J.B. Experimental nutrient additions accelerate terrestrial carbon loss from stream ecosystems. *Science* **2015**, *347*, 1142–1145. [CrossRef] [PubMed]

8. Neff, J.C.; ChaPin, F.S., III; Vituosek, P.M. Breaks in the cycle: Dissolved organic nitrogen in terrestrial ecosystems. *Front. Ecol. Environ.* **2003**, *1*, 205–211. [CrossRef]

9. Yuan, X.C.; Lin, W.S.; Pu, X.T.; Yang, Z.R.; Zheng, W.; Chen, Y.; Yang, Y. Effects of forest regeneration patterns on the quantity and chemical structure of soil solution dissolved organic carbon matter in subtropical forest. *Chin. J. Appl. Ecol.* **2016**, *27*, 1845–1852.

10. Galloway, J.N.; Townsend, A.R.; Erisman, J.W.; Bekunda, M.; Cai, Z.; Freney, J.R.; Martinelli, L.A.; Seitzinger, S.P.; Sutton, M.A. Transformation of the nitrogen cycle: recent trends, questions and Potential solutions. *Science* **2008**, *320*, 889–892. [CrossRef] [PubMed]

11. Nadelhoffer, K.J.; Emmett, B.A.; Gundersen, P.; Gundersen, P.; Kjønaas, O.J.; Koopmans, C.J. Nitrogen deposition makes a minor contribution to carbon sequestration in temperate forests. *Nature* **1999**, *398*, 145–148. [CrossRef]

12. Frey, S.D.; Ollinger, S.; Nadelhoffer, K.; Bowden, R.; Brzostek, E.; Burton, A.; Caldwell, B.A.; Crow, S.; Goodale, C.L.; Grandy, A.S.; et al. Chronic nitrogen additions suppress decomposition and sequester soil carbon in temperate forests. *Biogeochemistry* **2014**, *121*, 305–316. [CrossRef]

13. Findlay, S.E. Increased carbon transport in the Hudson River: Unexpected consequence of nitrogen deposition. *Front. Ecol. Environ.* **2005**, *3*, 133–137. [CrossRef]

14. Gundersen, P.; Emmettb, A.; Kjonaas, O.J.; Koopmans, C.J.; Tietema, A. Impact of nitrogen deposition on nitrogen cycling in forests: A synthesis of NITREX data. *For. Ecol. Manag.* **1998**, *101*, 37–55. [CrossRef]

15. Tu, L.; Hu, X.; Zhang, J.; Li, H.; He, Y.; Tian, X.; Xiao, Y.; Jing, J. Effects of simulated nitrogen deposition on soil active carbon pool and root biomass in Neosinocalamus affinis plantation, Rainy Area of West China. *Acta Ecol. Sin.* **2010**, *30*, 2286–2294.

16. Song, X.; Zhou, G.; Jiang, H.; Yu, S.; Fu, J.; Li, W.; Wang, W.; Ma, Z.; Peng, C. Carbon sequestration by Chinese bamboo forests and their ecological benefits: assessment of Potential, Problems, and future challenges. *Environ. Rev.* **2011**, *19*, 418–428. [CrossRef]

17. Song, X.; Chen, X.; Zhou, G.; Jiang, H.; Peng, C. Observed high and persistent carbon uptake by Moso bamboo forests and its response to environmental drivers. *Agric. For. Meteorol.* **2017**, *247*, 467–475. [CrossRef]

18. Zhou, G.; Jiang, P.; Xu, Q. *Carbon Fixing and Transition in the Ecosystem of Bamboo Stands*, 2nd ed.; Science Press: Beijing, China, 2010.

19. Song, X.Z.; Zhou, G.M.; Gu, H.H.; Li, Q. Management practices amplify the effects of N deposition on leaf litter decomposition of the Moso bamboo forest. *Plant Soil* **2015**, *395*, 391–400. [CrossRef]

20. Ma, S.; Li, Z.; Wang, G.; Liu, R.; Mao, F. Effects of intensive and extensive management on soil active organic carbon in bamboo forests of China. *Chin. J. Plant Ecol.* **2011**, *35*, 551–557. [CrossRef]

21. Li, Q.; Song, X.Z.; Gu, H.H.; Gao, F. Nitrogen deposition and management practices increase soil microbial biomass carbon but decrease diversity in Moso bamboo plantations. *Sci. Rep.* **2016**, *6*, 28235. [CrossRef] [PubMed]

22. Peng, C.J.; Li, Q.; Gu, H.H.; Song, X.Z. Effects of simulated nitrogen deposition and management type on soil enzyme activities in Moso bamboo forest. *Chin. J. Appl. Ecol.* **2017**, *28*, 423–429.

23. Song, X.; Li, Q.; Gu, H. Effect of nitrogen deposition and management practices on fine root decomposition in Moso bamboo plantations. *Plant Soil* **2017**, *410*, 207–215. [CrossRef]

24. Li, Y.F.; Jiang, P.K.; Liu, J.; Wang, X.D.; Wu, J.S.; Ye, G.; Zhou, G.M. Effect of fertilization on water-soluble organic C, N, and emission of greenhouse gases in the soil of *Phyllostachys edulis* stands. *Sci. Sil. Sin.* **2010**, *46*, 165–170.

25. Zhou, G.M.; Wu, J.S.; Jiang, P.K. Effects of different management models on carbon storage in *Phyllostachys Pubescens* forests. *J. Beijing For. Univ.* **2006**, *28*, 51–55.

26. Li, S.S.; Du, Y.X. Study advance on dissolved organic matter in soil. *Jiangxi For. Sci. Technol.* **2010**, *3*, 23–26.

27. Wu, J.S.; Jiang, P.K.; Chang, S.X.; Xu, Q.F.; Lin, Y. Dissolved soil organic carbon and nitrogen were affected by conversion of native forests to plantations in subtropical China. *Can. J. Soil Sci.* **2010**, *90*, 27–36. [CrossRef]

28. McClain, M.E.; Richey, J.E.; Brandes, J.A. Dissolved organic matter and terrestrial-lotic linkages in the central Amazon basin of Brazil. *Glob. Biogeochem. Cycles* **1997**, *11*, 295–311. [CrossRef]

29. Gielen, B.; Neirynck, J.; Luyssaert, S.; Janssens, I.A. The importance of dissolved organic carbon fluxes for the carbon balance of a temperate Scots pine forest. *Agric. For. Meteorol.* **2011**, *151*, 270–278. [CrossRef]

30. Jones, D.L.; Hughes, L.T.; Murphy, D.V.; Healey, J.R. Dissolved organic carbon and nitrogen dynamics in temperate coniferous forest plantations. *Eur. J. Soil Sci.* **2008**, *59*, 1038–1048. [CrossRef]

31. Yang, M.; Li, Y.F.; Li, Y.C.; Chang, S.X.; Yue, T.; Fu, W.J.; Jiang, P.K.; Zhou, G.M. Effects of Inorganic and Organic Fertilizers on Soil CO2 Efflux and Labile Organic Carbon Pools in an Intensively Managed Moso Bamboo (Phyllostachys Pubescens) Plantation in Subtropical China. *Commun. Soil. Sci. Plan Anal.* **2017**, *48*, 332–344. [CrossRef]

32. William, H.M.; Timothy, W. Podzolization: Soil processes control dissolved organic carbon concentrations in stream water. *Soil Sci.* **1984**, *137*, 23–32.

33. Vance, G.F.; David, M.B. Dissolved organic carbon and sulfate sorption by spodosol mineral horizons. *Soil Sci.* **1992**, *154*, 136–144. [CrossRef]

34. Pang, X.; Bao, W.; Wu, N. Influence Factors of Soil Dissoluble Organic Matter (Carbon) in Forest Ecosystems: A Review. *Chin. J. Appl. Environ. Biol.* **2009**, *15*, 390–398. [CrossRef]

35. Neff, J.C.; Asner, G.P. Dissolved organic carbon in terrestrial ecosystems: synthesis and a model. *Ecosystems* **2001**, *4*, 29–48. [CrossRef]

36. Mulholland, P.J.; Dahm, C.N.; David, M.B.; DiToro, D.M.; Fisher, T.R.; KögelKnabner, I.; Meybeck, M.H.; Meyer, J.L.; Sedell, J.R. What are the temporal and spatial variations of organic acids at the ecosystem level. *Fresh Water* **1989**, 315–329.

37. Jiang, P.K.; Xu, Q.F.; Xu, Z.H.; Cao, Z.H. Seasonal changes in soil labile organic carbon pools within a Phyllostachys praecox stand under high rate fertilization and winter mulch in subtropical China. *For. Ecol. Manag.* **2006**, *236*, 30–36. [CrossRef]

38. Treseder, K.K. Nitrogen additions and microbial biomass: A meta analysis of ecosystem studies. *Ecol. Lett.* **2008**, *11*, 1111–1120. [CrossRef] [PubMed]

39. Berg, B. Litter decomposition and organic matter turnover in northern forest soils. *For. Ecol. Manag.* **2000**, *133*, 13–22. [CrossRef]

40. Micks, P.; Downs, M.R.; Magill, A.H. Decomposition litter as a sink for ^{15}N-enriched additions to an oak forest and a red Pine Plantation. *For. Ecol. Manag.* **2004**, *196*, 71–87. [CrossRef]

41. Vitousek, P.M.; Howarth, R.W. Nitrogen limitation on land and in the sea: How can it occur? *Biogeochemistry* **1991**, *13*, 87–115. [CrossRef]

42. Aber, J.; McDowell, W.; Nadelhoffer, K.; Magill, A.; Berntson, G.; Kamakea, M.; McNulty, S.; Currie, W.; Rustad, L.; Fernandez, I. Nitrogen saturation in temperate forest ecosystems: Hypotheses revisited. *Bioscience* **1998**, *48*, 921–934. [CrossRef]

43. Bowman, W.D.; Cleveland, C.C.; Halada, L.; Hresko, J.; Baron, J.S. Negative impact of nitrogen deposition on soil buffering capacity. *Nat. Geosci.* **2008**, *1*, 767–770. [CrossRef]

44. Van DenBerg, L.J.; Peters, C.J.; Ashmore, M.R.; Roelofs, J.G. Reduced nitrogen has a greater effect than oxidised nitrogenon dry heathland vegetation. *Environ. Pollut.* **2008**, *154*, 359–369. [CrossRef] [PubMed]

45. Zhao, M.X.; Wang, W.Q.; Zhou, J.B. Effects of temperature on adsorption of soluble organic carbon and nitrogen from manure on arable soils. *Acta Pedol. Sin.* **2013**, *50*, 842–846.

MDPI
St. Alban-Anlage 66
4052 Basel
Switzerland
Tel. +41 61 683 77 34
Fax +41 61 302 89 18
www.mdpi.com

Forests Editorial Office
E-mail: forests@mdpi.com
www.mdpi.com/journal/forests